FIGHTING IN THE STREETS

FIGHTING IN THE STREETS

A Manual of Urban Guerilla Warfare

Urbano

Barricade Books Inc.
New Jersey

Published by Barricade Books Inc.
1530 Palisade Avenue
Fort Lee, NJ 07024
by agreement with J. Flores Publications

Distributed by Publishers Group West
4065 Hollis
Emeryville, CA 94608

Neither the author nor the publisher assumes any responsibility for the use or misuse of the information contained in this book. The author and publisher specifically disclaim any personal liability, loss, or risk incurred as a consequence of the use and application, either directly or indirectly, of any advice or information presented herein.

Technical data presented herein, particularly technical data on dealing with dangerous chemicals, poisons, explosives, drugs, and safety procedures inevitably reflects the author's individual beliefs and experience with particular equipment and environments under specific circumstances which the reader cannot duplicate or experience exactly. The information in this book should therefore be used for guidance only and should be approached with great caution.

Since neither the author nor the publisher have control over the materials used, use of techniques or equipment, or the abilities of the reader, no responsibility, either implied or expressed, is assumed for the use or misuse of the data or procedures given in this book.

Printed in the United States of America.

Library of Congress Cataloging-in-Publication Data

Urbano.
Fighting in the streets: a manual of urban guerilla warfare / Urbano.
p. cm.
ISBN 0-942637-47-X (pbk.) : $8.95
1. Guerilla warfare. 2. Urban warfare. I. Title.
U240.U73 1991
355.4'26—dc20 91-38512
CIP

WARNING!

Neither the author nor the publisher assumes any responsibility for the use or misuse of the information contained in this book. The author and publisher specifically disclaim any personal liability, loss, or risk incurred as a consequence of the use and application, either directly or indirectly, of any advice or information presented herein.

Technical data presented herein, particularly technical data on dealing with dangerous chemicals, poisons, explosives, drugs, and safety procedures inevitably reflects the author's individual beliefs and experience with particular equipment and environments under specific circumstances which the reader cannot duplicate or experience exactly. The information in this book should therefore be used for guidance only and should be approached with great caution.

Since neither the author nor the publisher have control over the materials used, use of techniques or equipment, or the abilities of the reader, no responsibility, either implied or expressed, is assumed for the use or misuse of the data or procedures given in this book.

CONTENTS

INTRODUCTION

Imagine that you are in the following situation: your country has been invaded by a foreign military power. Foreign troops occupy the cities and the capitol; a puppet government rules in the name of the invaders. Citizens have been subjected to mass searches and arrests. The invaders confiscate radios, weapons, batteries and anything else which may be used to challenge their domination. Spies, informers and undercover security agents are everywhere. Those who speak out, however innocently, against the repression disappear without a trace.

Another situation: your country is ruled by an oppressive and totalitarian dictatorship. The means of communication—press, radio, television—are controlled by the regime. The Constitution is suspended; elections are fraudulent or nonexistent. Freedoms are restricted. Critics of the regime vanish in the night.

In these situations, what would you do? Would you simply accept your position and carry on as best as you can? Would you hope and wish that the situation will somehow change of itself? Or will you take action and fight back, to overthrow the regime or to expel the invaders?

This manual is for these last persons, who are determined to fight for their freedom. It is for those ordinary citizens who, having decided that the regime must go, work to remove it in the only way that remains available to them—armed insurrection.

This manual is a description of the means which can be used to carry out that struggle. Since the military forces of the regime are strong and the military power of the population is weak, the resistance network must adopt the strategy of the guerrilla; lightning "hit and run" attacks, striking repeatedly and vanishing, leaving the regime stung, bewildered and demoralized.

The majority of guerrilla wars have been fought in isolated, rugged areas, in mountains or jungles which provide safe sanctuaries for the fighters. In modern industrialized states such as the US, the Soviet Union or Europe, such a rural-based strategy will fail. Since the regime's military, economic and political resources are concentrated in the cities, the resistance fighters must be "urban guerrillas," carrying out their fight in the cement jungles of the urbanized areas. It is through urban warfare that the regime can be toppled.

1

PRINCIPLES OF URBAN GUERRILLA WARFARE

Guerrilla warfare is the natural method of struggle against a powerful enemy. It is obvious that the forces of the occupied population are not, at least initially, able to stand in face-to-face combat against the professional armed forces of the regime or against the occupation forces of the invader. For this reason, the guerrilla adopts the strategy of nibbling at the edges of the enemy, of harassing and tiring him rather than confronting him directly. Guerrilla warfare is above all the strategy of "hit and run," of "fighting and running away to live to fight another day." It is a warfare of deception and illusion, of surprise and the unexpected.

In undeveloped agricultural nations such as China, Cuba, El Salvador, the Philippines, Afghanistan or the central Asian Soviet Republics, the primary focus of the guerrilla fight is in the countryside. These nations have conditions in the rural areas which are suited for a protracted guerrilla war: rough inaccessible terrain, few avenues of transportation, a thin population distribution, and little or no system of communication. In these types of territory, rural guerrilla warfare is the most effective method of fighting the regime.

In the industrialized urban nations of North America, Europe or the Soviet Union, such a strategy is unworkable. In these societies, guerrillas must operate in the urban nerve centers to attack the occupying regime.

The primary characteristics of a successful rural guerrilla war are: mobility, surprise, local superiority in forces, intimate knowledge of the terrain, and the support of the population. These tactics have been described in numerous texts on rural guerrilla warfare.

On the whole, urban guerrilla warfare tactics do not differ in any significant way from those used by rural guerrillas. The resistance fighter who combats an industrialized, urban oppressor must depend on the same guerrilla principles of mobility, surprise, local superiority, knowledge of the terrain and the support of the population.

Nevertheless, the urban guerrilla faces problems which his rural counterpart does not. The urban fighter operates in the very center of the regime's power. The rural guerrilla can slip away from a fight knowing it may take days or even weeks for reinforcements to arrive on the scene: the urban guerrilla may be located and surrounded by the regime's police and troops in a matter of minutes. The rural fighters operate in mountains and jungles far away from prying eyes: the urban guerrilla lives and operates in the midst of large crowds of people, any one of whom may be an informer or collaborator. Thus the security problems of the urban fighter are compounded tremendously.

The urban fighter does, however, possess some advantages. His actions and successes cannot be hid from the people by the regime's censorship, since they are immediately visible. This gives hope to the population and demonstrates that the regime is not unbeatable. More concretely, the urban fighter will find that resources such as food, shelter, clothing, weapons and ammunition are much more readily available in the city than they would be in the mountains or jungle. Also, the urban resistance fighter will find that the masses of people provide a shield into which he can melt, allowing him to remain anonymous until he strikes.

The targets of the urban guerrilla differ somewhat from those of the rural fighters. While the rural units attack the enemy's troop outposts and supply convoys, the urban guerrilla attacks such centralized targets as banks and financial centers, communications hubs, power stations, military and police stations, arsenals, government offices and officials of the regime.

Mobility is an important weapon for the urban guerrilla, since he can be so easily and rapidly surrounded by the state security forces. Such an encirclement can prove fatal to the fighters. For this reason, the urban guerrilla's actions must be meticulously planned so the guerrilla team is able to slip away before the state security forces arrive.

The automobile is an invaluable aid to the urban guerrilla, and every fighter must be familiar with methods of commandeering or expropriating cars, vans, and trucks. Often, a single operation will involve the procurement of several different vehicles.

The urban guerrilla must carry the concept of "hit and run" to the extreme. The fighters arrive at the target, launch their attack, and vanish into the streets, all in a matter of minutes.

Since the vehicles used in the operation may be described to the state security forces, the resistance fighters must make provisions to evade pursuers and roadblocks which will hinder the escape. A switch of vehicles shortly after the operation will confuse pursuers. Monitoring the radio frequencies used by the state security services will tell the fighters of roadblocks or other obstacles. The urban fighter must take every step to avoid being encircled and forced into a firefight.

The urban guerrilla must also be thoroughly familiar with the terrain in which he operates. He must know every street, every alley, the location of empty or half-built structures, and even the layout of sewers and subterranean tunnel systems. The resistance fighters must be able to make their way quickly through a maze of obstacles while the state security forces stumble blindly behind them, ready to fall into a fatal trap or ambush. Pursuers must be slowed with mines, grenades, caltrops or snipers stationed along the escape route.

Because the urban guerrilla forces are so small compared to the repressive forces available to the dictator or invader, the tactic of surprise is an urgent necessity in urban operations. Most operations will take place so quickly that the enemy will have no time to react. Guards and state security forces which are found in the area of the operation must be quickly disarmed or eliminated. The guerrilla attack must be swift, sudden and unexpected.

For this reason, the ambush and the raid are the urban guerrilla fighter's stock in trade. Rather than attacking the regime's forces head-on and falling to his superior firepower, the guerrilla chooses moments when he has an unbeatable advantage before swiftly pouncing on the enemy, wiping him out and vanishing before reinforcements can arrive.

Because the guerrilla force as a whole is hopelessly outnumbered, no battles should be accepted by the fighters until they have a superiority of force in the area of conflict. This is accomplished by forcing the enemy to divide his forces among many dispersed targets. By attacking at various places, the resistance network forces the enemy to garrison troops at all such potential targets as supply depots, bridges, power stations, etc., and thus to spread his forces too thinly.

A guerrilla force of forty fighters, for instance, is helpless against an army of a thousand. But if these thousand men are divided among fifty potential targets, the guerrilla force need face only twenty men at each target, giving them local superiority of force. The enemy is never able to concentrate his forces since he never knows when or where the guerrillas will strike next. He is thus forced to guard all potential targets or else leave a number of targets unguarded and vulnerable.

The support of the population is the most important weapon of the guerrilla. People who support and sympathize with the resistance provide the logistical aid that the guerrillas need, including food, sanctuary for wounded fighters, new recruits and scouts in unfamiliar terrain.

The local population is also vital to the resistance fighters in providing intelligence concerning the movements of enemy forces. Local civilians serve as the eyes and ears of the guerrilla forces, informing them of enemy troop movements in the area, details of potential targets for guerrilla raids, and other pertinent information. At the same time, a sympathetic population serves as a shield for the resistance network by refusing to give information about the movement to the enemy, or by giving false information which actually aids the resistance. The enemy is thus constantly watched by thousands of eyes and his every move known to the guerrillas, while the guerrillas themselves remain invisible to the regime's armed forces.

At the same time, the urban guerrilla must realize that he is very vulnerable to infiltration or betrayal, and that a single civilian could cripple the resistance network greatly by passing information to the authorities. The urban guerrilla must be selective in his dealings with civilians, and his security system must be leakproof if he is to survive. The urban guerrillas, like any other guerrillas, still must swim like fishes, but now their water amounts to a mere puddle.

The most common operations carried out by urban guerrillas are raids on arsenals and medical supply houses to obtain needed weapons and materials, raids to release imprisoned guerrillas, kidnappings or executions of occupation authorities, and raids on installations such as airfields, military and police bases, power stations, and factories which are important to the regime (weapons factories, heavy industrial plants, power stations). Other actions include the armed support and protection of civilian rallies and mass demonstrations against the regime and the use of armed raids and sabotage as a tactical adjunct to strikes, boycotts and work slowdowns which cripple the regime.

Initially, the guerrilla movement will consist of scattered detachments which operate in a single city or area. As the resistance grows, these link up and combine operations. Coordinated strikes will paralyze the enemy's control over a region or area of the country. These regional struggles grow in time to assume the character of a full-scale war, with the guerrilla forces operating more and more as a regular army.

While this army begins a general offensive against the enemy, the civilian population fights with actions such as mass demonstrations and general strikes, which cripple and weaken the regime. The final step is the mass uprising which sweeps the hated regime away.

2

THE URBAN BASE OF OPERATIONS

In order for the guerrilla force to elude pursuers and survive to carry out new attacks, it must maintain a base of operations in which to maintain equipment, tend to wounded, and rest between operations. For the urban guerrilla fighter, this base takes the form of the urban "safehouse."

Note that this base of operations is not by any means a "rear area" or "safe retreat." Even in an area which is relatively secure, to remain encamped in the same spot for longer than a few days is to invite a fatal encirclement by the state security forces. Even during periods between operations, the guerrilla safehouse is moved constantly, eluding the regime's efforts to locate and destroy it. It must be, like the guerrillas themselves, small and highly mobile.

Two types of safehouse are required by the urban resistance network, the "barracks" and the "operations center." The barracks serve as temporary shelters for guerrillas, where they can eat and sleep in safety. The operations centers are temporary arsenals, weapons factories or other clandestine centers where the tasks of the resistance can be carried out without interference from the state.

Guerrillas who are unknown to the police can live undercover in "normal" houses, and have no need of a guerrilla barracks. Fighters and network members who have been identified by the state security forces, though, must carry on their work completely "underground," with no known addresses or fixed places where they can be captured. These clandestine fighters must be provided with secure safehouses which are completely unknown to the authorities, where they can safely stay between guerrilla operations. They must be billeted in barracks which are placed at the disposal of the resistance network by sympathetic members of the population.

Operations such as arms factories, arsenals or supply depots must also be located in urban safehouses which are unknown to the authorities and are untraceable. These secure areas are also provided by sympathetic supporters of the resistance.

The security needs of both types of safehouses are similar. Safehouses must be nondescript and unobtrusive, indistinguishable from neighboring houses. Nothing about the house or its occupants should hint or remotely suggest that it is anything other than a typical neighborhood home. To allay suspicions, the occupants of the house must not appear to depart from normal routine.

Urban guerrillas should do nothing to suggest that they are anything other than ordinary "law-abiding citizens." The dress, habits, actions, language and lifestyles of the guerrillas should conform completely to the area in which the house is located. If the safehouse is in a poor working class area, for example, it would be suicidal for the guerrillas to wear business suits and drive expensive cars. Similarly, guerrillas billeted in middle class suburban areas must not draw unnecessary attention to themselves by wearing old clothes, having long hair, driving old cars, etc. Urban guerrillas and their safehouses must melt inconspicuously into the local population.

The urban safehouse must have a good view of the surrounding area and provide good fields of fire over approaching streets and alleys. Window shades or blinds should have slits cut into them for inconspicuous observation. A lookout must be constantly maintained to prevent the security forces from surrounding the urban camp unnoticed.

Houses which are isolated or surrounded by open areas are unsuitable for urban bases, since they can be easily surrounded and covered by a few armed security forces. Row homes are best, since the security forces must occupy the whole block in order to surround the house. The security forces will also be forced to enter and search every house in the area for escaped guerrillas and stragglers, a time-consuming and dangerous process which exposes them to snipers and guerrilla raids.

The safehouse should be surrounded by a security net to prevent an encirclement by the police and also to watch for enemy agents who will try to surveil the suspected hideout. Resistance agents posing as street workers, people in restaurants, people at a phone booth, or people going out for a walk will circle the area of the safehouse, watching for enemy agents or police forces. If a suspected agent approaches, the security net immediately contacts the safehouse using coded clandestine communications, giving the illegal occupants time to hide or escape and allowing the legal "front" to reassure the police agents. If the security forces approach in force, the security network warns the safehouse so that it can be evacuated quickly, leaving the police with nothing.

When the safehouse is first occupied, at least two emergency escape routes should be reconnoitered and rehearsed. In the event of a raid, the urban guerrillas seek to hold off the enemy as long as possible, allowing resistance members to remove equipment and supplies and melt away singly or in pairs to evade the police net and disappear in the crowds of bystanders and passersby. At least one member of the safehouse should be armed with a rifle or submachine gun, positioned so he can cover the escape of his fellows.

If the safehouse cannot be defended and important materials must be left behind, the guerrillas should arm a pre-set booby trap which will be detonated when the security forces enter the building, destroying the supplies to prevent the enemy from gaining intelligence from them. A booby trap at the house's entrance will delay the police, allowing more time for the guerrillas' escape.

The urban guerrilla must also take great care in entering or leaving the urban base. Any excursion outside the safehouse, for however short a period, must be considered as a military operation, and must be planned and thought-out. Before leaving the safehouse, watch carefully for suspicious persons or vehicles who might be state security agents. Once on the street, be as inconspicuous as possible. Keep a sharp eye out for state security men who may be following you. Police tails must be lost. The easiest way to do this is to use public transportation during peak traffic hours, when it is easy to become swallowed up in the faceless crowd of people. Underground resistance fighters should spend as little time as possible in public, leaving the safehouse only to carry out a task or operation. Tasks such as buying food, collecting mail, and so on, must be performed by the legal "front person" who occupies the safehouse.

Whenever possible, resistance fighters should appear to lead a normal life, with a family, job, etc., to avoid drawing suspicion from the police. Safehouses and barracks are only for those guerrillas who have been identified by the state security forces and are forced to operate underground. A few of these "wanted" fighters may choose to operate with the rural guerrillas, where they can remain out of the reach of the police, rather than continue to risk capture in the city. For those who decide to continue urban work, secure safehouses are a necessity.

The legal "front" people who obtain the resistance's safehouses should be above suspicion and with no known connections to the guerrilla movement. Safehouses should be obtained by members of the population who are completely unknown to the police, with no known rebel sympathies and no suspicious political connections. If the police raid the safehouse, this "legal" front must be capable of answering the police's questions in a manner that will satisfy them and remove the house from suspicion.

Urban safehouses should be loaned from these front people, not rented. No guerrilla or guerrilla sympathizer should rent a safehouse, either under a legal identity or an illegal one. This is done to avoid the possibility of the state security forces tracing this paper trail to a member of the resistance network. Urban camps must be borrowed or loaned from legal "aboveground" people with no known connections to the guerrillas. If a temporary base is needed for the purpose of a specific operation (for instance a vantage point from which to provide fire support for a raid on an arsenal or police station) the guerrillas must occupy a suitable building and prevent bystanders from interfering until the operation is complete.

The urban safehouse must also be, like the urban guerrillas themselves, free-floating and mobile. No urban base should be occupied for longer than a week or two, to prevent security forces of the regime from tracking down the guerrillas and raiding them. To depend on a small number of fixed safehouses is to hinder the mobility and flexibility of the fighters, and to establish a fixed front is to make the job of the police easier. No fixed urban bases should be maintained or defended. Rather, the guerrillas should develop a free-floating, mobile network of safe bases among a sea of supportive people. It thus becomes an impossible task for the regime's police to find the constantly-moving urban bases. As with the rural base camp, the urban base depends on its unobtrusiveness and its mobility for its security, not on its defensive firepower. At all costs, urban guerrillas must avoid being surrounded and trapped in their safehouses.

Arms factories, arsenals, printing shops, and other urban bases which cannot be easily moved must be set up in a remote area which is absolutely unsuspected. Disguise and concealment are a necessity. Cached weapons, explosives, and other supplies may be cleverly hidden in the house or its surroundings, while the legitimate front person works to allay police suspicions.

The guerrilla's network of safehouses is vital to the success of the resistance. By developing the ability to vanish into an unknown hideout among a sea of sympathetic people, the guerrillas can escape the state security forces at will, vanishing after an operation and emerging somewhere else to carry on the fight. The regime comes to feel as though it is chasing a rabbit which has an unlimited number of tunnels in its warren. The regime never knows where the guerrillas will appear next.

acid.

Chlorate/sugar igniter is most easily utilized in the form of a solid mass known as "fire fudge." This is formed simply by dissolving one part granulated sugar in hot water, removing from heat and dissolving one part of potassium chlorate powder. This forms a sticky syrup which solidifies upon cooling into a hard white substance that looks like candy. This solid substance will burst into flame in contact with acid or glycerin, and is also set afire by a safety fuse.

The most common filling for incendiary grenades and bombs is napalm, which is simply gelled gasoline. Napalm burns fiercely and clings to whatever it hits.

Napalm can be improvised by the guerrilla armorer in several easy ways. In the easiest method, mix 36 parts by volume of gasoline with 1 part of 100-proof alcohol (whiskey or vodka) and 25 parts of soap flakes. Only real soap can be used, such as Ivory Soap, Lux Soap Flakes or Palmolive Soap Bars.

Detergent will not work. Put the gasoline in a bucket or other container and add the alcohol. Stir the soap flakes in slowly until the gasoline sets in a thick gel. After standing for a few days, the mixture will have the consistency of butter or vegetable shortening. If necessary, it can be thinned a little by adding more gasoline.

Gasoline can also be gelled with egg whites and any of the following additives: instant coffee, table sugar, Epsom salts, baking soda, table salt, or crushed aspirins. To make napalm, place the gasoline in a bucket and add 1 part of egg whites (without the yolks) to every six parts of gasoline. Slowly add coffee, sugar or some other suitable material until the gasoline gels to the consistency of jam. This version of napalm breaks down rather quickly and should be used within 24 hours.

An incendiary mixture useful for fire grenades is made from potassium chlorate, granulated sugar, sulphur, iron filings and wax. Use a double boiler to melt the wax, sulphur, sugar and iron together in a small amount of water. The proportions should be one and a half parts sulphur, two parts sugar, one part iron filings and one and a half part candle wax.

Remove this mixture from heat and carefully add 4 parts potassium chlorate powder to the mixture. Stir well to mix it together. Be careful at this point; the chlorate is liable to sparks and friction until the mass has cooled. This incendiary mixture can be ignited by the heat from a small piece of "fire fudge."

While napalm incendiaries are most effective against enemy personnel, and the chlorate mixture described above is useful against wooden structures, incendiary attacks against metal machinery require a special incendiary. The military uses a mixture called Thermite which burns at over a thousand degrees, and this mixture can be easily improvised by the home armorer. Thermite is nothing more than a mixture of three parts iron oxide (ordinary rust) with two parts of powdered aluminum.

Thermite is not ignited by most other incendiary mixtures, and a special igniter and booster is needed for reliable ignition. The igniter is formed by mixing aluminum powder, sulphur and starch into a solid mass. Mix 6 tablespoons of aluminum powder and 15 tablespoons of sulphur together. In a separate pot, dissolve two tablespoons of starch to 6 ounces of water and boil gently. When the starch is dissolved, cool the solution and add the aluminum/sulphur powder. Mix this into a thick paste and form it into little cylinders an inch and a half long by an inch wide. Before it is dried, punch a hole in the top for a safety fuse to fit inside.

The booster for Thermite is made with a mixture of one part aluminum powder and one part of potassium nitrate or potassium permanganate. This booster will ignite Thermite mixture, and is itself easily ignited by the sulphur/aluminum pellet and fuse.

Using the incendiary, chemical and explosive compounds described in this chapter, the resistance network will be able to manufacture a variety of improvised weapons for use until the guerrillas are strong enough to liberate their weapons and supplies from the state security forces.

4

IMPROVISED WEAPONS AND MUNITIONS

In the beginning of a resistance struggle, the guerrilla fighters will have limited resources with which to fight. Weapons and ammunition must be captured wherever possible from the armed forces of the regime to increase the firepower and fighting ability of the guerrilla forces.

In the early stages of the movement, the fighters will lack these basic arms and must therefore make do with improvised and easily constructed substitutes. In some circumstances, a resistance unit may find itself isolated and lacking in fighting weapons. Under these circumstances, the ability to improvise effective weapons is vital.

This chapter, therefore, will concern itself with the manufacture of effective military weapons from commonly available materials. Together with the improvised explosives and chemicals described earlier, it will provide the means for the resistance movement to carry out attacks and liberate needed supplies, materiel and weapons.

FUSES AND DETONATORS

Fuses are necessary for igniting an improvised incendiary charge or in detonating a blasting cap for a demolition charge. Commercial fuse can be captured from industrial sources or expropriated from mines, quarries or construction sites.

A workable improvised fuse can be produced quickly by soaking a length of string or shoelace in a strong solution of potassium nitrate and allowing it to dry. To decrease the burning rate of this improvised fuse, soak string in a 50-50 mix of granulated sugar

MATCHBOOK IGNITER

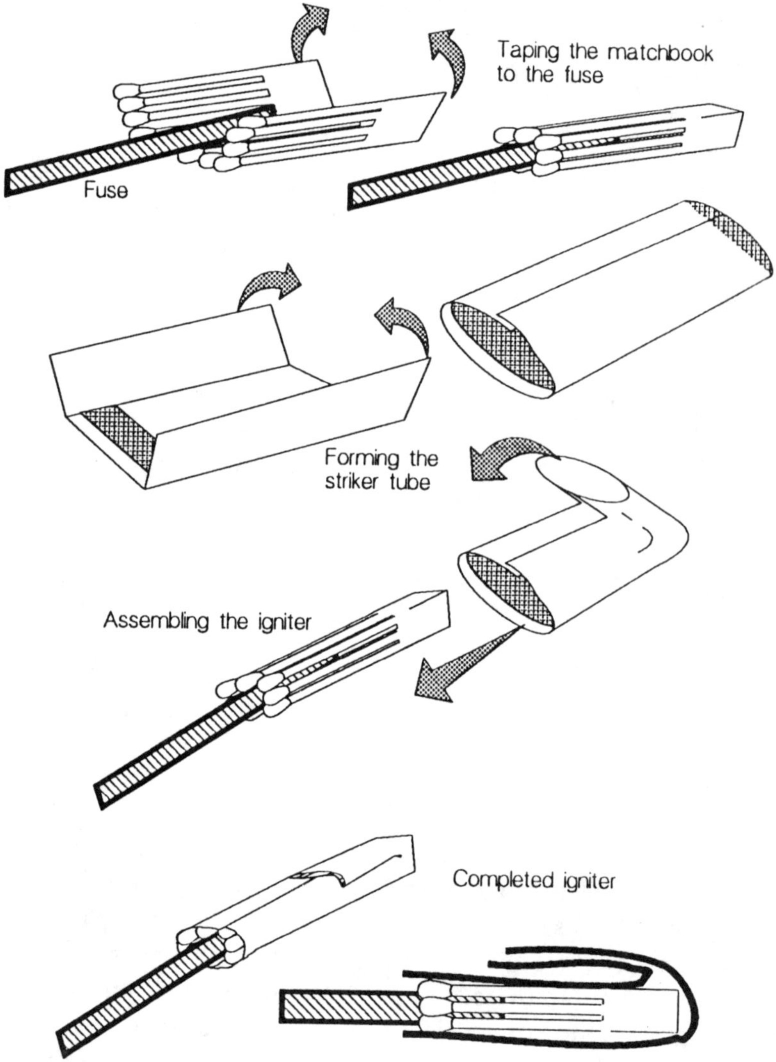

and potassium nitrate. Braid three of these treated strands together to produce a crude fuse. Test to determine the burning rate of each batch of fuse produced.

A much more reliable and workable fuse can be produced with only a little more effort. Mix a quantity of black powder with a little glue and enough water to form a thin paste. Soak cotton string or shoelaces in this mixture so the thinned gunpowder paste adheres to it. For thicker fuses, braid two or three strands of sting together before dipping. When the gunpowder paste has dried, wrap the improvised fuse in waterproof masking tape. Test a length of fuse to determine the burning time and then cut the fuse strand into the desired length.

Safety fuse is difficult to light with matches or lighters, and therefore some method must be improvised for reliable fuse ignition. A good fuse igniter can be improvised using an ordinary paper book of matches.

Separate the paper matches from the cover of the matchbook. Cut one end of the fuse so a large area of the powder inside is exposed and tape it securely to the paper matches so the end of the fuse is at the center of the row of match heads.

Fold the row of matches over the fuse and tape them down.

Next, shape the matchbook cover into a tube that will fit snugly over the paper matches. Make sure that the striking area of the matchbook cover tube is on the inside. Slide this tube over the match/fuse assembly until the striking surface is slightly below the match heads. A portion of the matchbook cover tube should be projecting beyond the fuse. Fold this over and tape it down.

To prevent the matchbook cover from working its way off the igniter, pin it in place with a small needle or safety pin.

To use the igniter, remove the safety pin and grasp the matchbook cover tube firmly. Pull it off of the igniter with a swift, firm motion. This causes the match heads to rub against the striking surface, lighting the matches and producing a flame which ignites the fuse.

This igniter cannot be used in wet or damp weather. It also produces a bright flame at the point of ignition, which may not be desirable during clandestine operations.

An all-weather fuse igniter which remains invisible during use can be produced with homemade chemicals and a few pipe fixtures.

Cut a half-inch length of solid steel bar which has been threaded on the outside. In one end, drill a hole just big enough for the safety fuse to fit inside snugly. Drill this hole almost, but not quite, completely through the piece of steel. At the other end, drill a one-sixteenth inch hole which leads into the fuse hole.

Find a pipe cap which fits over this threaded piece of steel and drill a small hole in the closed end. Fit a short nail with a large flat head through this hole with the flat head inside the cap. Cover the bottom of the pipe cap with two layers of strike-anywhere match tips. Do not use the entire match head—just cut off the initiating tips. Screw the loaded pipe cap over the piece of threaded steel until it is finger tight. Be careful not to crush the match tips. Waterproof the igniter by pouring melted candle wax around all joints.

The igniter is triggered by sharply tapping the point of the nail on a rock or other hard surface. This crushes the match tips and sets them on fire, which in turn ignites the fuse.

A more reliable igniter is built on the same principles, but uses the reaction of sulfuric acid and potassium chlorate to produce flame. Instead of match tips, fill the pipe cap with tissue or paper towel which has been soaked in a strong potassium chlorate or powdered match head solution. Take a length of glass tubing which fits snugly inside the pipe cap and seal one end with wax. Fill this tube three-fourths of the way with concentrated sulfuric acid and seal the tube closed. Place the chlorate paper and the acid tube in the pipe cap and screw this onto the threaded piece of steel, taking care not to break the glass tube.

To use the igniter, tap the point of the nail sharply onto a hard surface. The flat head of the nail will crush the glass tube, exposing the chlorate paper to acid and causing it to burst into flame.

An effective blasting cap for triggering improvised explosives can be made with a length of fuse, improvised low explosive, waterproof masking tape and a paper or cardboard tube about three-eights inch wide and three or four inches long. Crimp the

NON-SPARKING FUSE IGNITERS

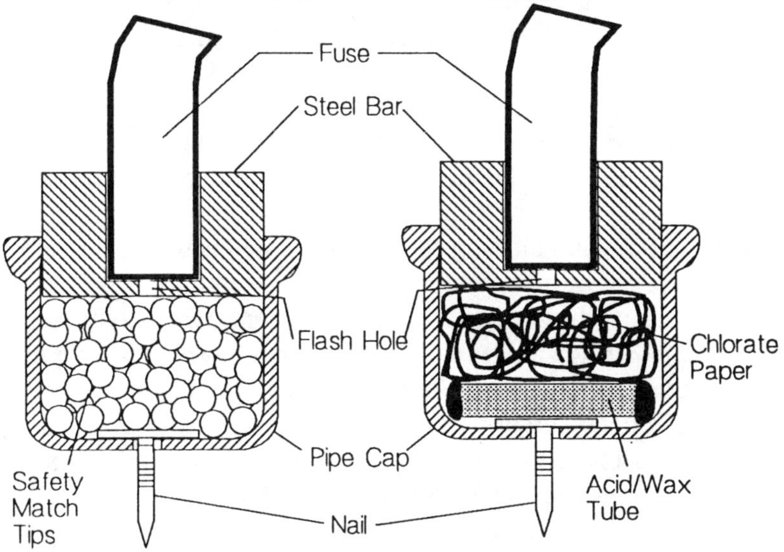

IMPROVISED BLASTING CAPS

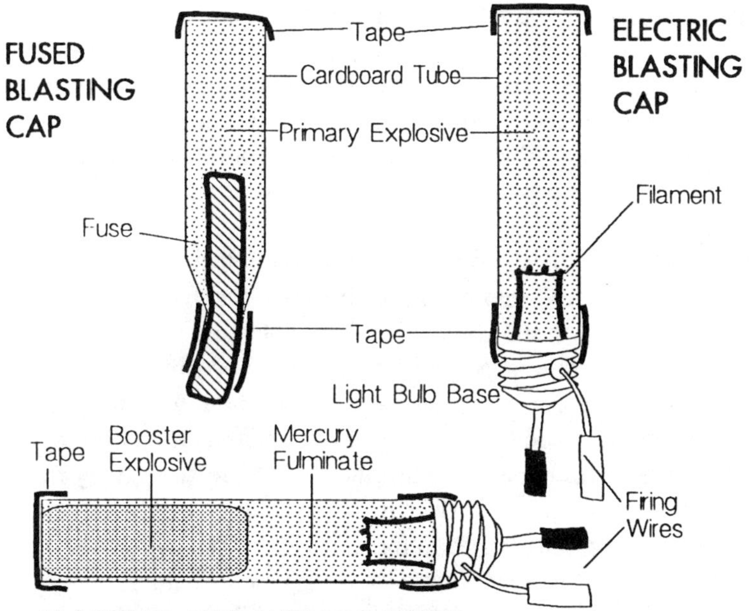

BLASTING CAP WITH BOOSTER

tube around the safety fuse, taking care that at least an inch of the fuse projects into the tube, and tape it in place. Fill the tube with an 80-20 mixture of mercury fulminate and potassium chlorate. If these are unavailable, black powder can be used instead. Tape the other end of the tube closed. This blasting cap will detonate all burster charges and low explosives. It will not detonate insensitive explosives such as ammonium nitrate, dynamite or RDX plastique.

A more powerful blasting cap capable of detonating these high explosives can be made with the same basic setup.

Make the cardboard tube about an inch longer, and fill the fuse end with a half gram of fulminate/chlorate or black powder. Fill the rest of the tube with picric acid, gelatin dynamite or ammonium nitrate/mercury fulminate booster. Wrap the booster explosive in a small plastic bag before inserting, to prevent contact between the booster and the primary explosive. This blasting cap should be capable of detonating any improvised explosive charge.

The same blasting cap may also be converted to fire electrically for use in demolitions and command-detonated mines. Break the glass from a flashlight bulb, taking care not to break the filament of the bulb. Solder or glue two firing wires to the base of the bulb and fit this into one end of a cardboard or paper tube so the filament projects inside. Tape the tube securely in place around the bulb base. Fill the tube with a gram and a half of primary explosive, or one-half gram of primary explosive and two or three grams of booster.

To use, connect the two firing wires to an electric circuit and close the switch. Current passing through the bulb heats the filament, setting off the powder inside the blasting cap.

The current from four penlight batteries or two ordinary flashlight batteries, wired in series, should be enough to heat the filament and detonate the blasting cap. If necessary, an ordinary bicycle headlight generator, the kind which are rotated by the wheel to produce electricity, can be used as a firing circuit. Connect the two firing wires to the generator—one wire goes on the terminal and the other is connected anywhere on the generator case. Running the generator drive wheel rapidly across the palm of the hand will produce a burst of current strong enough to ignite the blasting cap.

Short-term batteries can also be improvised to power electrical firing circuits. The guerrilla will need:

One copper plate measuring about 4 x 4 x 1/16 inch
One aluminum plate, same size
Powdered charcoal, potassium permanganate crystals or man ganese dioxide
Sodium hydroxide (soda lye) crystals

Mix one part of the soda lye crystals in two parts water. If liquid soda lye can be procured, it is acceptable if it contains at least 45% by weight lye. Place this solution in a leakproof container.

Thoroughly mix the powdered charcoal and the potassium permanganate or manganese dioxide powder. Do not grind these together—just mix them. Manganese dioxide can be obtained from the inside of dead flashlight batteries. Mix this powder with water to form a thick paste.

Spread a layer of this paste on the copper plate, about a quarter of an inch thick. Be careful not to get this solution on exposed skin.

A few minutes prior to use, pour enough of the soda lye solution on the charcoal paste to moisten all of it. Press the aluminum metal plate on top of this, taking care that no portion of the aluminum plate comes into contact with the copper plate. Tape one of the firing lines to the aluminum plate and the other to the copper plate. When the circuit switch is closed, the crude battery will deliver enough current to detonate one blasting cap. This battery will go dead in about ten minutes or after one firing.

A stronger improvised battery can be manufactured with many of the same materials. The guerrilla will need:

Three aluminum plates, same size as before.
Two copper plates, same size.
Charcoal powder.
Manganese dioxide powder or potassium permanganate.
Sal ammoniac (ammonium chloride).

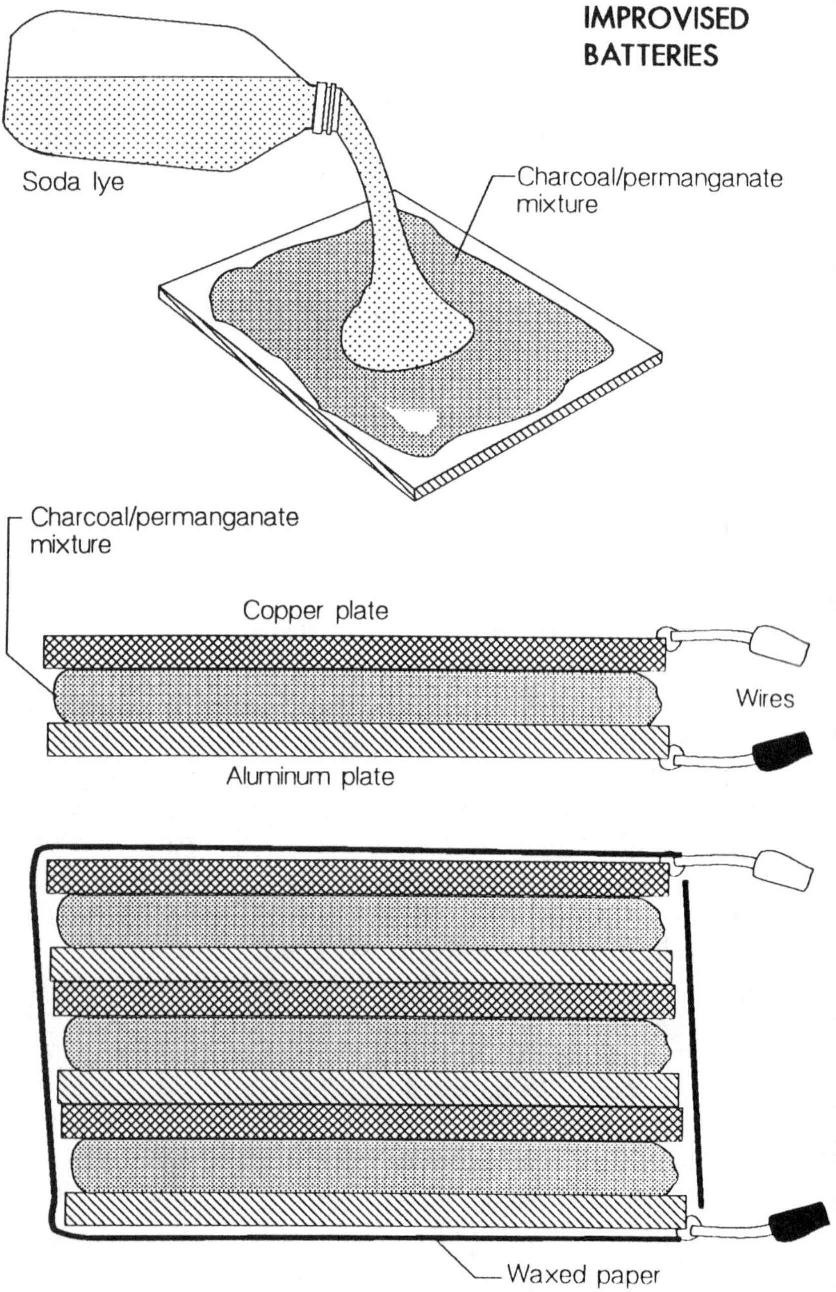
IMPROVISED
BATTERIES
Soda lye
Charcoal/permanganate mixture
Charcoal/permanganate mixture
Copper plate
Wires
Aluminum plate
Waxed paper

Mix the sal ammoniac with water to form a solution of about 45% strength. Sal ammoniac is found in some soldering fluxes and fertilizers. Mix one part of sal ammoniac solution with one part charcoal and one part permanganate or manganese dioxide. Spread a thick layer of this paste onto the copper plate and press the aluminum plate on top, making sure that the two plates do not touch. Prepare three of these cells and pile them on top of each other, with the copper plate of one cell resting on top of the aluminum plate of the cell below it. Wrap this whole package in heavy waxed paper to protect it from moisture. Expose the top and bottom metal plates (one will be copper, one aluminum) at one corner. These are the terminals of the batteries. Tape the firing lines to these corners.

A three-cell unit provides enough current to fire one blasting cap at a time. The battery will produce current for about two hours and is good for three or four firings.

In many instances, it is necessary for the demolition charge or incendiary device to have a built-in delay period to allow the guerrilla to escape the area before the device is activated. Improvised delay systems can self-activate the device for periods from six or seven minutes to ten or more hours after it is set at the target.

One common method of delay depends on the chemical reaction of concentrated sulfuric acid and potassium chlorate to ignite the device. Place an end of the safety fuse in a plastic tube and pack a 50-50 mixture of potassium chlorate and granulated sugar around it. Seal one end of a glass tube which will fit inside the plastic tube, using paraffin or wax. Fill this three-fourths of the way with concentrated sulfuric acid. Cover the open end of the acid tube with a piece of rubber balloon or a rubber condom and seal it tightly with a rubber band or glue. This seal must be absolutely leakproof.

To activate the device, place the glass acid tube inside the plastic chlorate tube so the rubber barrier rests on top of the chlorate/sugar mixture, which in turn rests on top of the fuse end. It will take about six or seven minutes for the acid to eat through the rubber barrier. When it contacts the potassium chlorate mixture, it will

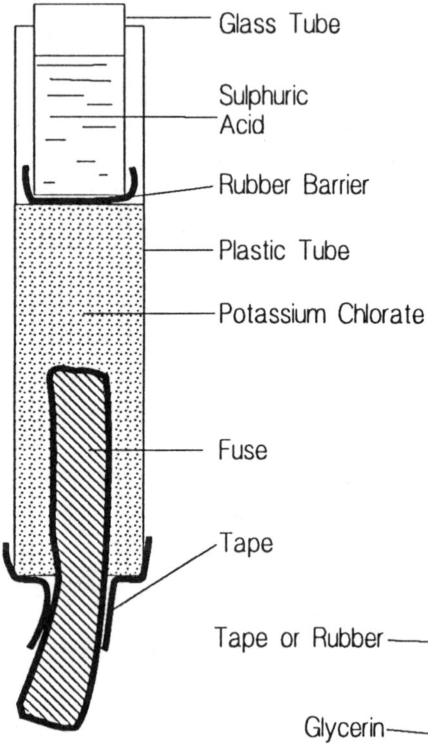

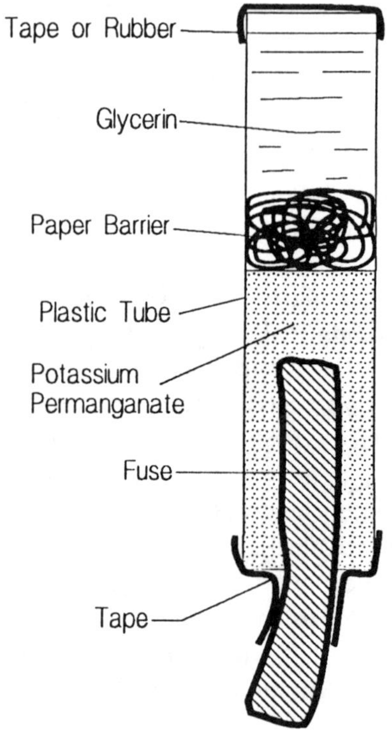

CHEMICAL
DELAY
DEVICES

produce a flame which will ignite the fuse and set off the device. Before the device is actually used, a sample should be tested to determine the exact delay time.

To increase the delay time, use two or more layers of rubber barrier.

A similar device uses glycerin and potassium permanganate to produce flame. Fit the end of the fuse into a plastic tube and pile a half-inch or so of potassium permanganate crystals on top of it. Stuff a plug of tissues or paper towel on top of this to form a barrier about half inch thick. Carry separately a small bottle of glycerin.

To activate the delay mechanism, pour an inch or so of glycerin on top of the paper barrier. Depending on the thickness and solidity of the barrier, the glycerin will soak through in five or six minutes. When it reaches the permanganate crystals, it produces flames which ignite the fuse. The delay time is adjusted by adding more or less paper barrier to the trigger.

More precise and longer delay times can be produced using an improvised electric timer to close the firing circuit at a predetermined time. This type of delay works well with demolition charges using electric blasting caps rather than safety fuse. A workable timer can be made with an ordinary wristwatch with a glass or plastic face.

Allow the watch to run down completely before beginning assembly. The watch must have metal hands and a glass or plastic crystal face. If the watch has a moving second hand, remove it, and if the desired delay time is more than 59 minutes, remove the minute hand also. If there is paint on the watch hands, scrape it off to insure a good contact.

Drill a small hole through the watch crystal and insert a small screw or nail. This must fit tightly and must not touch the face of the watch. The hand of the watch must not be able to pass this screw without touching it.

Solder or glue a firing wire to the top of this screw. Then solder or glue the other firing wire to the metal portion of the watch case. Insulate these wires so they do not form a complete circuit.

Set the hands of the watch so that one hand will touch the screw at the desired time. Then wind the watch and connect the firing

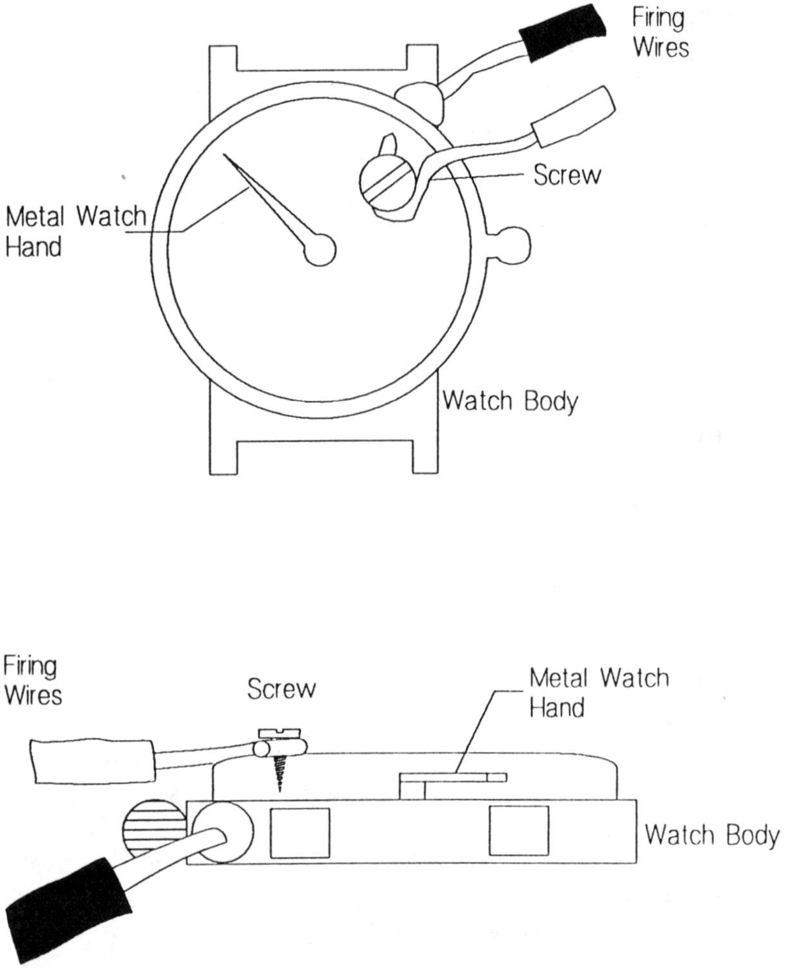

IMPROVISED ELECTRIC TIMER

lines to the circuit. When the hands of the watch touch the screw, it completes the circuit and detonates the charge.

MINES, BOMBS AND DEMOLITION CHARGES

Charges used for demolition and sabotage depend for their effect on the force of the explosion rather than on fragmentation or shrapnel effects. These charges can take the form of bare chemical explosives which are placed on the object or structure to be demolished.

The main charge for demolition should be a high explosive such as gelatin dynamite, ammonium nitrate or some other powerful explosive. To use dynamite or ammonium nitrate, wrap the explosive in paper tubes to form sticks. Tape several of these sticks tightly together to form a concentrated charge. Use a wooden stick to poke a hole in the central stick of dynamite and insert the detonator/booster blasting cap. Make sure this fits snugly and will not fall out. Detonate by lighting the fuse or closing the electrical firing circuit.

Since the blast effect of an explosive does not travel easily through air, the charge should be pressed as tightly against the target as possible. The force of the explosion will be more directly communicated to the target if a thin layer of wet mud is plastered between the explosive charge and the target. It is also a good idea to place the blasting cap in the side of the main charge which is furthest from the target.

Demolition using plastique charges is more efficient because the plastique can be molded tightly around the object to be demolished, increasing the effect of the explosion. To demolish a target using plastique, simply mold a large lump of plastic RDX or potassium chlorate around the target and insert a blasting cap/booster. Detonate by cord or by electricity.

These explosive charges will serve to demolish wooden structures, delicate machinery, vehicles, and other soft targets. Hardened steel targets such as heavy machinery, locomotive engines, armored vehicles, etc., must be attacked using improvised concentrated explosive known as "shaped charges." These explosive arrangements produce a high-velocity jet of gases and metal which can puncture steel plate and armor and destroy it.

An improvised shaped charge requires an amount of plastic or other explosive, a blasting cap, a thin metal sheet and a metal or cardboard tube about a foot long and three or four inches wide.

Cut the metal sheet into a circle and bend it into a cone which will fit tightly into the metal or cardboard tube. Place the cone so the pointed end is facing up, about four or five inches from one end. Pack the upper part of the tube with plastique or other explosive. Finally, insert a blasting cap/booster assembly.

The tube provides the proper "stand-off" distance for the explosive jet to develop. The device is placed on the object to be pierced and detonated. The metal cone will disintegrate and form a molten plasma which will pierce the metal armor and destroy it.

If no metal is available for the cone, a shaped charge can be improvised with a glass soda bottle. Simply fit a tube over the pointed end of a soda bottle and fill it with explosive. The bottom half of the bottle will provide the proper stand-off distance for the charge.

The simplest type of fragmentation anti-personnel bomb is the crude pipe bomb. This simply consists of a length of steel pipe, a foot or so long and at least an inch wide, which is threaded at both ends for a pipe cap. Screw one pipe cap on tightly and fill the pipe with a bursting charge such as black powder, flash powder, or guncotton. More powerful bombs can be made with ammonium nitrate or shredded dynamite. Insert a blasting cap into the explosive charge and screw on the other cap, after having drilled the necessary holes through this cap for the blasting cap wires or fuse. To increase the fragmentation effect, wrap the pipe with several turns of heavy steel wire, or, before filling it with burster, grind shallow depressions in a checkerboard pattern into the pipe. A timing device can be taped to the outside of the pipe. It is also a good idea to tape or wire the pipe bomb to a flat piece of board to prevent it from rolling.

The explosive charge should be wrapped in a plastic bag before being placed in the pipe, to prevent a chemical reaction with the metal.

Self-activated pressure mines can be improvised using a metal

IMPROVISED SHAPED CHARGES

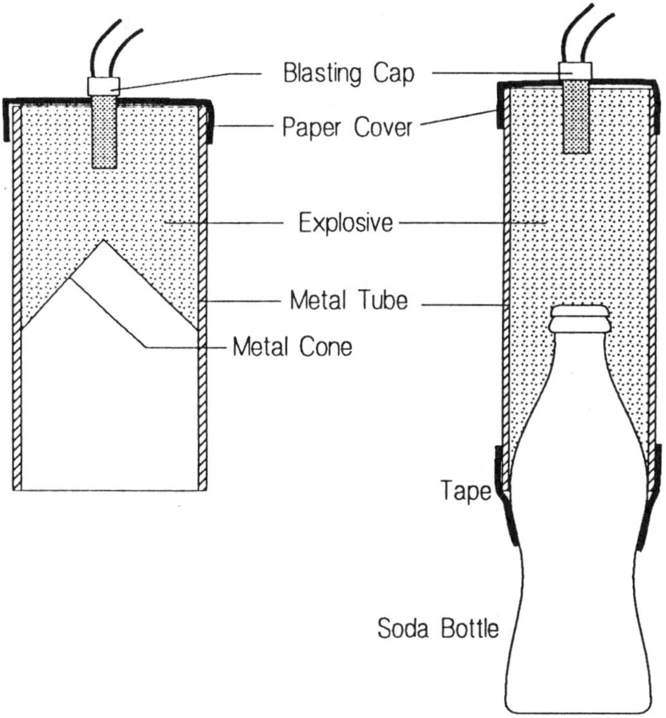

PIPE BOMBS

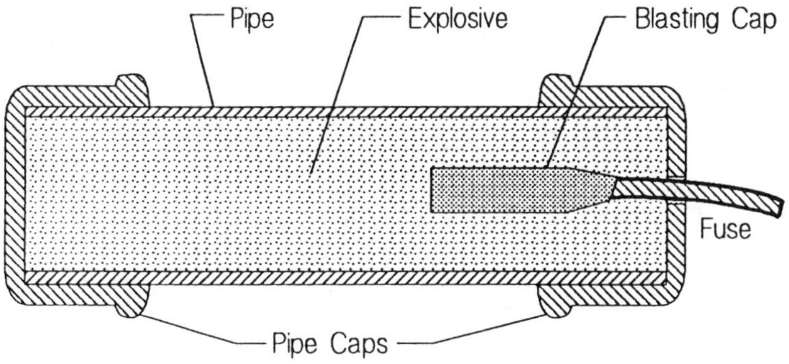

IMPROVISED ANTI-VEHICLE MINE

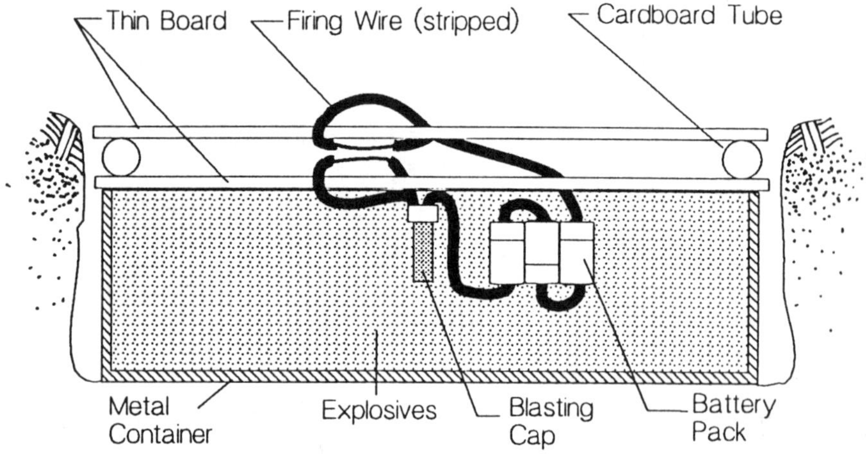

IMPROVISED ANTI-PERSONNEL MINE

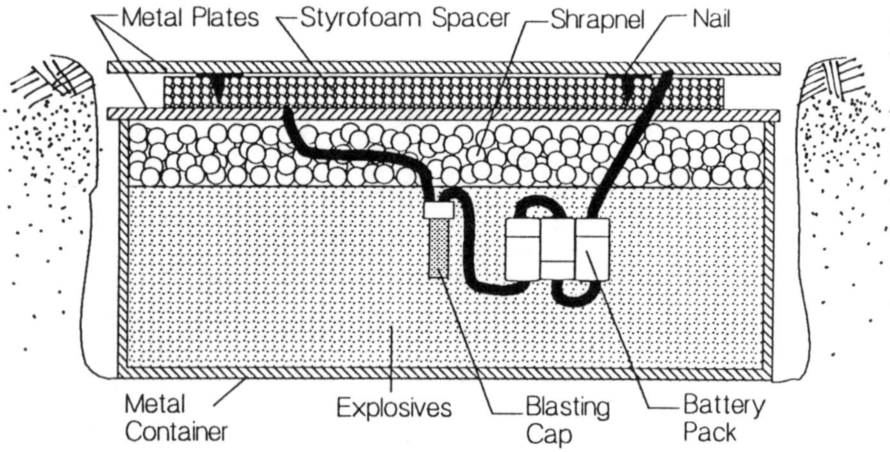

or cardboard container, a blasting cap, a main explosive charge, a couple of flashlight batteries, some wire, two pieces of cardboard or wood, and two cardboard toilet paper or paper towel rolls.

The main explosive is packed into the bottom of the container and a blasting cap is inserted. In anti-vehicle pressure mines, this charge should consist of high explosives such as dynamite or ammonium nitrate. For anti-personnel fragmentation mines, a charge of flash powder or black powder will do, but it must be covered with two inches of BBs, small nuts or bolts, small rocks, nails, or some other material to serve as shrapnel.

A flat piece of wood is placed over the explosive charge, and a wire is wrapped several turns around this. The insulation is stripped from the wire along the middle portion of the top of the board. This wire is run to one terminal of the battery pack through the blasting cap. The other piece of board is similarly wrapped with wire which has been stripped in the middle, and this board is wired to the other battery terminal. The two wired boards are held apart by the toilet paper tubes, and the whole apparatus is buried shallowly in a road of footpatch.

Pressure on the wooden boards causes the cardboard tubes to collapse, allowing the stripped wires to contact each other and closing the firing circuit.

A similar pressure mine can be improvised with a metal container, a flat piece of styrofoam, two metal plates and a few nails.

Fill the metal container with high explosives or a burster and shrapnel, and insert a blasting cap and battery pack.

Cover this with a piece of flat metal which has been wired to the blasting cap and battery at one corner. Insert a number of sharpened on a piece of styrofoam so they are almost exiting through the other side. Wire a corner of a second sheet of metal to the batteries and place on top of the mine, keeping the two plates separated by the styrofoam spacer. Bury the device shallowly.

Pressure on the mine will push the nails through the styrofoam, connecting the two wired plates and setting off the charge.

Stake mines can at times have advantages over the buried anti-personnel mines because they can be quickly and easily emplaced. An improvised stake mine can be made with a soup can, a battery/detonator, a wooden stake and a spring-type clothespin.

Fill a cardboard tube with flash powder or other burster and place it in the center of the soup can, leaving about an inch of space on all sides. Fill this space with BBs, nails, or other shrapnel. Insert the blasting cap and glue the lid of the can back on, after drilling holes for the firing wires. Tape the battery pack to the side of the stake. Below this, fasten a spring-type clothespin with a firing wire stripped and wrapped around each of the jaws of the pin, in such a manner that the wires will touch when the clothespin is closed.

Hold the jaws of the clothespin apart with a piece of plastic, wood or other insulating material. Fasten a thin trip wire to this insulator and string it across a footpath or trail at ankle height, where it is hidden by vegetation. Tie the other end to a tree or bush.

Enemy personnel who walk along the path will catch the trip wire with their feet and pull it, dislodging the insulator from the clothespin switch. The jaws close, connecting the firing wires and detonating the stake mine.

This same clothespin trigger can be used to set off various kinds of booby traps and bombs. Fasten the clothespin switch to a solid object and wire it to the detonator, which is inserted into a demolition charge of plastique or high explosive. The trip wire can be attached to a car door, drawer, or any other object which, when moved, pulls the insulator from the switch and sets off the bomb.

Another booby trap can be constructed using a modified pipe bomb. Fill the bomb with burster or high explosive and a booster quantity of black powder. Instead of a blasting cap and wires, insert a half-gram of wet ammonium tri-iodide into the bomb and screw on the cap. Tape the pipe to a board to prevent it from rolling and set it in place. After ten or fifteen minutes, the ammonium tri-iodide will dry, and the slightest vibration or motion will set it off, in turn detonating the bomb. Once set, this device cannot be removed without setting it off.

One of the disadvantages of the clothespin-triggered booby trap is that the device can be disarmed if it is discovered, simply by cutting the trip wire. An improvised trigger using a hacksaw blade,

IMPROVISED STAKE MINE

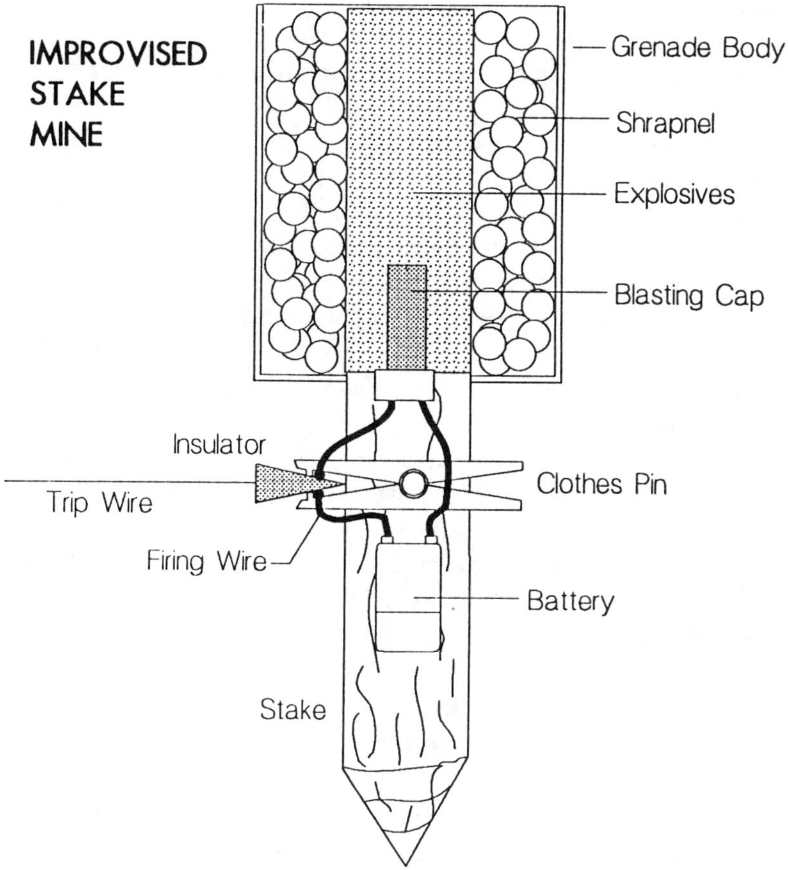

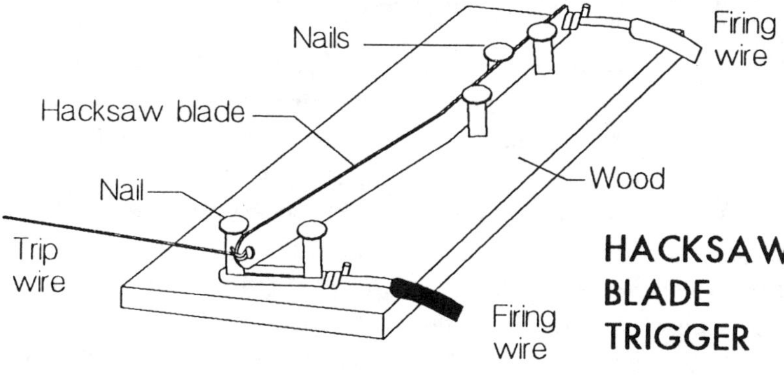

HACKSAW BLADE TRIGGER

however, will eliminate this danger.

The guerrilla will need a piece of wood, a few nails, and a flexible piece of metal such as a hacksaw blade or a thin knife blade. Nail one end of the hacksaw blade firmly to the wooden board at one end. Wire one of the firing lines to the metal at this end. At the other end, drive a nail into the board at a point where the hacksaw blade is firmly touching it, and drive another nail about an inch away so the piece of metal lies between them. Run a single firing line through both nails, taking care that none of the wires touch the hacksaw blade. Attach the trip wire so that its pressure holds the end of the hacksaw blade between the two nails. Then attach the firing lines to the battery/detonator pack.

If the trip wire is pulled, it will pull the hacksaw blade into contact with one of the nails, closing the circuit. If, on the other hand, the trip wire is cut, the relaxed hacksaw blade will contact the other nail and also set off the device. Once set, this device is very difficult to safely disarm. Be very careful when setting this device, however, as it is easy to accidentally trigger the booby trap while placing it.

A booby trap which can strike large areas at once is improvised with a cannon or mortar made from pipe. Use a section of steel pipe several inches wide and at least 18 inches long, with a pipe cap at one end. Fill the bottom of the pipe with several inches of black powder or other propellant, and insert a weak blasting cap into this.

Run the firing lines outside and screw on the pipe cap. Fill the rest of the pipe with shrapnel. Aim the device along a footpath or trail.

The trip wire is strung across the path, using either the clothespin or the hacksaw blade trigger. When the trip wire is pulled, it closes the electrical circuit and fires the cannon at the approaching enemy, spraying the area with flying shrapnel.

The claymore directed mine is another easily improvised weapon which is excellent for ambushes or for perimeter defense. Claymores can be improvised with a piece of sheet metal, some plastique, and some BBs or other shrapnel.

Cut the metal sheet into a rectangle measuring about 8 x 10

inches, with two pointed metal feet. Bend it into a flat arc. Press a charge of plastique onto the concave side of the metal, and insert a blasting cap into the center of the charge. On the convex side, glue several layers of BBs, nail or other shrapnel.

Claymores can also be improvised using a shallow metal pan or tray. Fill the pan to within a half inch with plastique or burster, and cover with a layer of paper. Fill the rest of the pan or tray with BBs or nails which have been coated with a layer of glue to hold them in. Punch a hole in the back of the pan to allow the detonator to be inserted into the charge.

The claymore is usually a command-detonated weapon which requires an operator to close the switch at the proper time. It can also be rigged to a trip wire as a booby trap or perimeter defense.

GRENADES

Grenades are the guerrilla fighter's equivalent of artillery. They are simply small hand-delivered bomblets which are filled with shrapnel, explosive, incendiary, chemical or other payloads.

The simplest type of grenade is the incendiary bottle, known popularly as the Molotov cocktail. This is used to start fires in buildings, or as incendiary weapons against vehicles or troops. It is nothing more than a glass bottle or jar which has been filled with gasoline and plugged with a gas-soaked rag which serves as a wick. When the rag is set aflame and the bottle is thrown, the gasoline is ignited and spreads a sheet of flame.

More effective Molotov cocktails can be made by using homemade napalm as a filler instead of gasoline. This burns hotter and will cling to whatever it splatters on.

One disadvantage of the Molotov cocktail is the fact that, since it must be lit before being thrown, it tends to give away the position of the guerrilla who is using them. This difficulty can be overcome with the use of impact-triggered incendiary bottles which do not light until they have reached their target.

The fighter will need a glass bottle or jar, a mixture of one part gasoline with one part motor oil, paper towels, potassium chlorate/sugar mixture, and concentrated sulfuric acid. Fill the bottle about three-fourths full with the gasoline/oil mixture. Then slowly add several inches of sulfuric acid to the mixture. Soak the paper towel in strong chlorate/sugar solution and wrap this securely around the bottle.

On impact, the bottle will break and spill its contents onto the chlorate paper. The sulfuric acid in the mixture will react with the paper to form a flame, which will ignite the gasoline. This triggering method is not reliable with napalm. The oil in the mixture is to allow the burning gasoline to cling to surrounding objects.

Since glass bottles are awkward and cannot be carried in quantity, guerrilla raiding parties should be equipped with small incendiary grenades which are capable of setting fire to selected targets.

The most critical part of any improvised grenade is the fuse and detonator assembly. Since all of the hand grenades described here use the same basic fuse assembly, we will consider the construction of this in some detail.

Grenade detonators are made with an aluminum tube, a wooden dowel which fits snugly into the tube, a length of fuse with a four or five second burning time, a large cotter pin, glass tubing, paraffin wax, concentrated sulfuric acid and a wad of cotton or paper towel which has been impregnated with a strong potassium chlorate/sugar solution.

The illustration demonstrates how the device is assembled. The wooden dowel is cut into two pieces. The glass tube is filled with sulfuric acid and sealed with wax. The fuse is threaded through a hole bored through the bottom half of the dowel and is sealed with wax or glue. The fuse/dowel is placed in the bottom of the aluminum tube and the chlorate paper is wadded on top, making sure that it contacts the end of the fuse. The acid tube fits snugly through a hole drilled through the aluminum tube, and the top half of the dowel is placed in the tube above it. Drill another hole through both the tube and the dowel to pin it in place with the cotter pin. The striker must not contact the acid tube.

The fuse/detonator is now ready. To use, remove the cotter pin to arm the grenade. Give a sharp blow to the striker dowel with the palm of the hand. This breaks the glass tube and spills acid onto the chlorate paper, igniting it. The fuse burns through the dowel and, after a five second delay, reaches the filler in the

IMPROVISED GRENADE FUSE

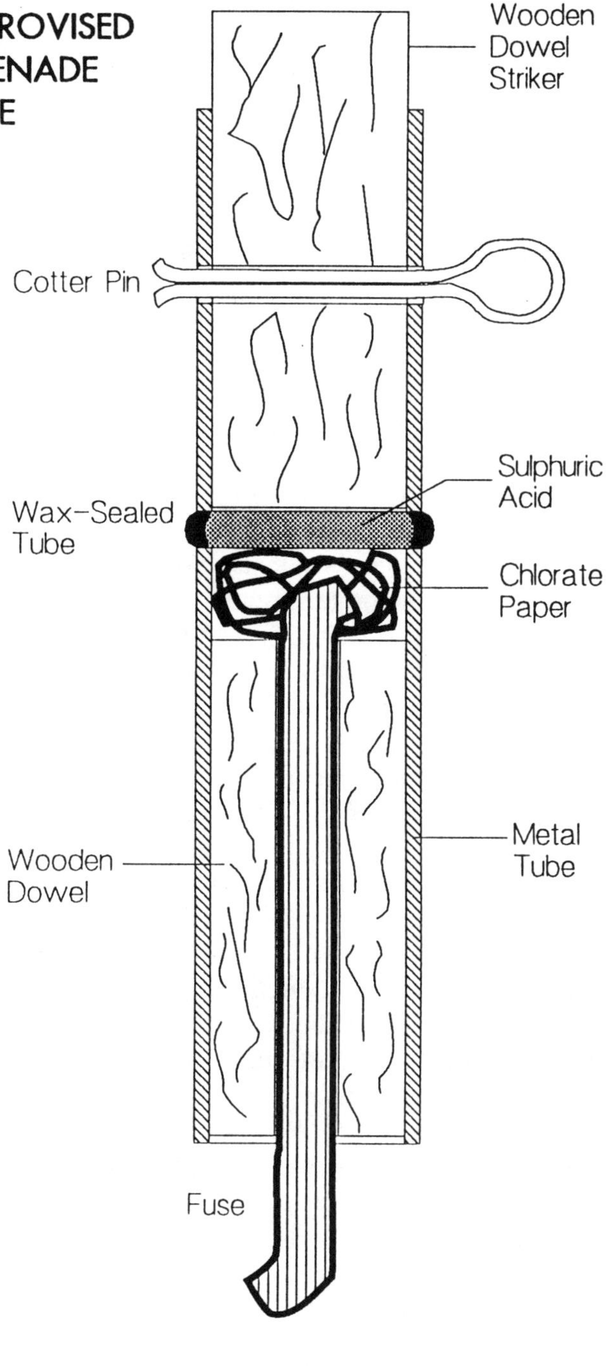

grenade body.

The grenade body of the incendiary grenade is made from a four-inch length of two-inch-diameter cardboard tube. A cardboard circle is glued to the bottom of the grenade body, and a similar circle with a hole punched in the middle is fitted over the fuse/detonator assembly.

The body of the grenade can be filled with improvised napalm. To prevent the gasoline from soaking through the cardboard tubing, the napalm is confined in a plastic bag.

A more efficient incendiary filler can be made from aluminum powder and potassium nitrate or potassium permanganate. A 50-50 mixture of aluminum powder and either of these oxidizers is poured into the grenade body, leaving a cavity in the middle for the fuse assembly.

This mixture is not reliably ignited by fuse, so an initiating pellet is made of sulphur and powdered aluminum. Mix these with water in the proportion of 15 parts sulphur to 6 parts aluminum to form a thick paste. Form this paste into balls about 3/4 inch in diameter. Use a pencil to poke a hole into the ball for the fuse to fit. Ignition can be made more certain by sprinkling a bit of chlorate/sugar powder into this hole before inserting a fuse.

The incendiary grenade is assembled as shown. The burning fuse ignites the sulphur/aluminum pellet, which produces enough heat to ignite the incendiary mixture. This burns fiercely with enough heat to ignite wood, paper or other fuels.

To attack metal machinery, the guerrilla can use a grenade containing the incendiary Thermite. This substance burns at over 1000 degrees, and can be used to warp precision machinery, weld delicate parts together or to burn through armor plate.

The grenade body for the Thermite grenade is an aluminum soup or beer can with the top cut off. Pack a paper tube with the incendiary mixture used in the incendiary grenade and place this in the center of the grenade body. The fuse/detonator extends into this tube. Fill the surrounding space with Thermite mixture. Thermite is composed of three parts rust and two parts aluminum powder. Cut a hole in a circular piece of metal and glue this in place over the fuse/detonator.

IMPROVISED INCENDIARY GRENADE

IMPROVISED THERMITE GRENADE

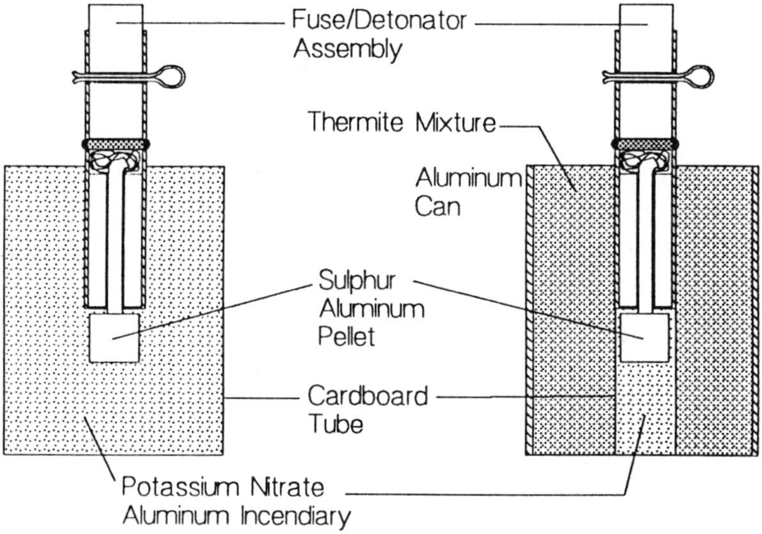

IMPROVISED CHEMICAL AND SMOKE GRENADE

FRAGMENTATION GRENADE

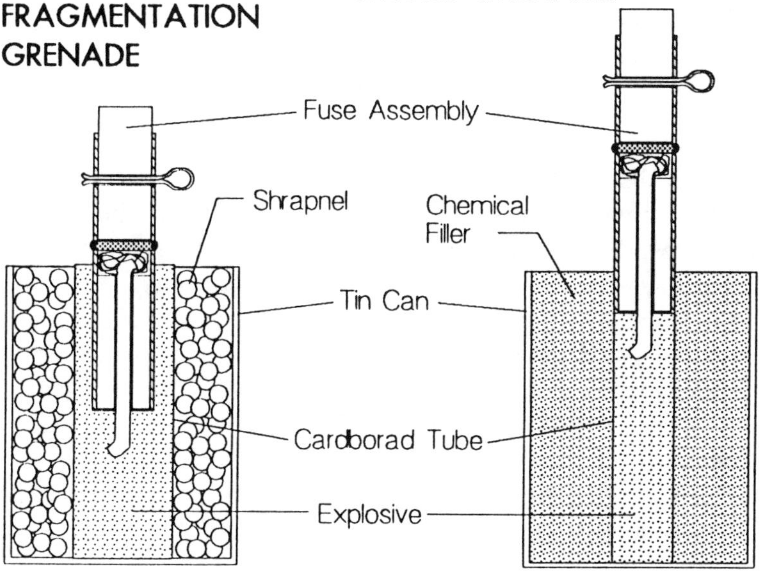

The fuse reaches the sulphur/aluminum pellet and the incendiary powder which ignites and produces enough heat to ignite the Thermite.

To burn holes through armor plate or machinery, place the grenade on a metal stand so it is suspended about four inches above the metal to be attacked. Arm the grenade. When the Thermite ignites, it will drip molten metal onto the armor plate and burn through it in a short time.

To weld parts together and ruin them, place the grenade body directly on the parts to be attacked. The heat from the Thermite will also heat the metal parts, allowing the molten metal to form an effective weld which will destroy the attacked machinery.

Fragmentation grenades are made with black powder, flash powder or chlorate/sugar as a burster. Pack the burster into a cardboard or paper tube and put it in the center of a cardboard grenade body. A tin can could also be used as a grenade shell. Surround the burster with BBs, nails or other shrapnel.

Fragmentation grenades can also be made from four inch lengths of pipe nipple, one inch in diameter, which are threaded on both ends for pipe caps. Cap the pipe at one end and fill with burster. To increase the fragmentation effect, grind a checkerboard series of depressions into the pipe before filling, or wrap the pipe with steel wire. Drill a hole in the second pipe cap to fit over the fuse/detonator and screw it on tightly.

In close quarters combat, it may be advantageous to the guerrillas to have grenades which will stun and disorient opponents without the danger of flying shrapnel. These stun grenades are made from cardboard tube in the same way as incendiary grenades. They are filled with flash powder and produce a brilliant flash, loud report and sharp concussion on detonation.

Smoke and chemical grenades may also be produced using the standard fuse/detonator assembly. Tin cans or soda cans can serve as grenade bodies. Workable smoke grenades can be made by filling the grenade body with used motor oil, which burns when ignited with a thick black smoke.

Chemical grenades use a small bursting charge of black powder or chlorate/sugar, enclosed in a small plastic bag, to burst open the grenade casing and vaporize the chemical to form a small cloud. Improvised tear gas grenades use acrolein or chloracetone as a filler. Lethal grenades use hydrogen cyanide.

All of these grenades can be modified in such a way as to be fired from 12-gauge shotguns as improvised grenade launchers. The body and fuse system of these launched grenades are the same as before, but the upper dowel striker consists of a long stick which reaches completely down the barrel of an improvised or sawed-off shotgun. The shotgun barrel is cut down to fourteen or fifteen inches. A rubber disc is attached to the butt of this stick so it fits the bore of the barrel snugly.

The propellant for the improvised grenade launcher is a 12-gauge shotgun shell with all of the shot and half the propellant powder removed. To use the weapon, load the shotgun and insert the grenade stick all the way into the barrel. Firing the shotgun drives the grenade from the barrel, and also pushes the striker into the acid tube, arming the grenade. The shell explodes five seconds after the weapon is fired. Maximum range is about three hundred yards.

Another weapon which can be improvised with easily-obtained equipment is the rocket-propelled grenade. RPGs are made with a five foot length of aluminum tube measuring two inches in diameter, four D size flashlight batteries and a tube which will hold these in series, a quantity of alligator clips and wire, and an electric firing switch used for toy model rockets. Projectiles for the launcher are made from model rocket engines and igniters, cardboard tube, stiff cardboard cards and sheet metal.

The rocket launcher is assembled as in the diagram.

The last six or seven inches of the top portion of the launching tube is removed for easy reloading. Two handles are mounted on the tube, and the firing switch is mounted on one of these. Firing cables run from the battery pack to a pair of alligator clips which are hooked to the rocket at the time of firing. The battery pack consists of four D size batteries placed end to end in a plastic tube, wired at each end. This pack is mounted on the rear of the launcher.

Rockets consist of cardboard tubes with D size or larger model rocket engines and igniters mounted at the rear. All of the rocket

IMPROVISED GRENADE LAUNCHER

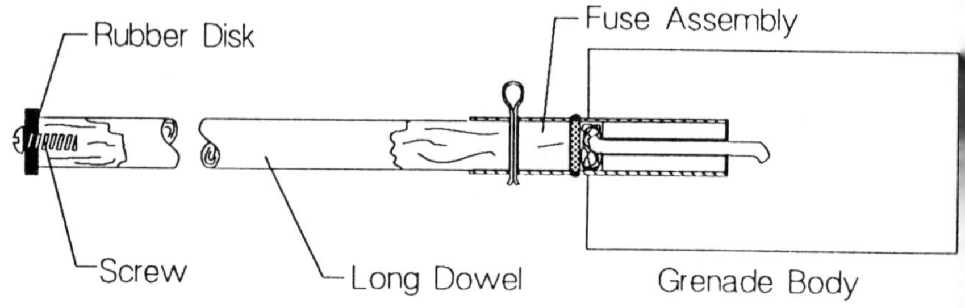

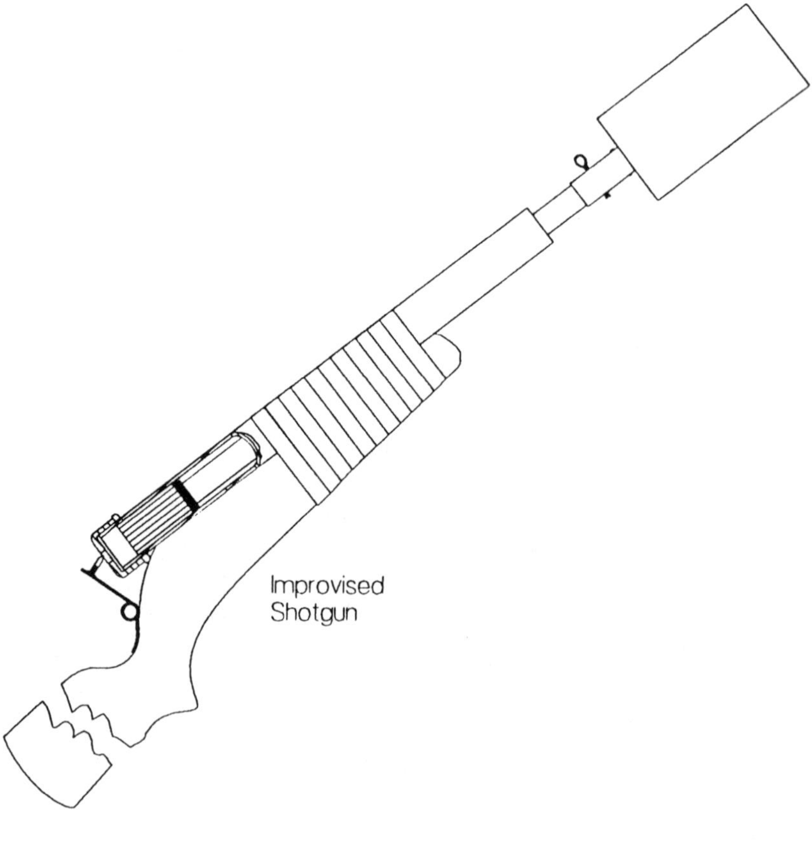

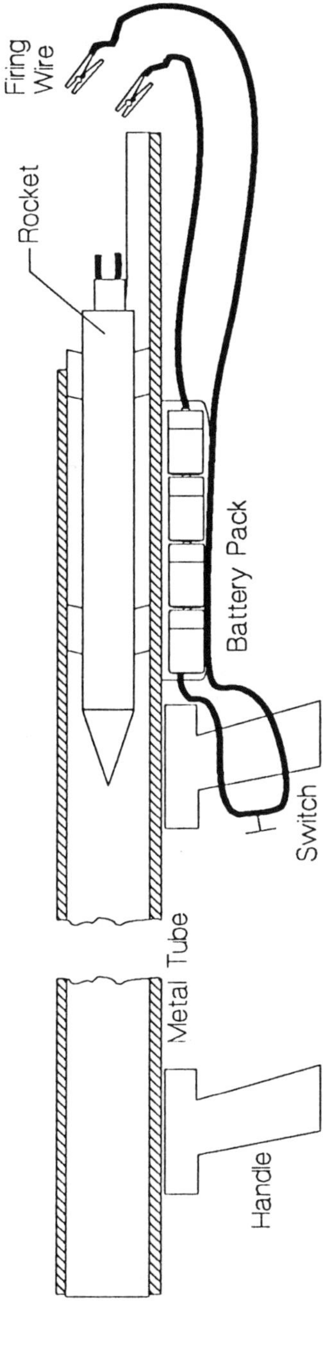

ROCKET LAUNCHER AND ROCKETS

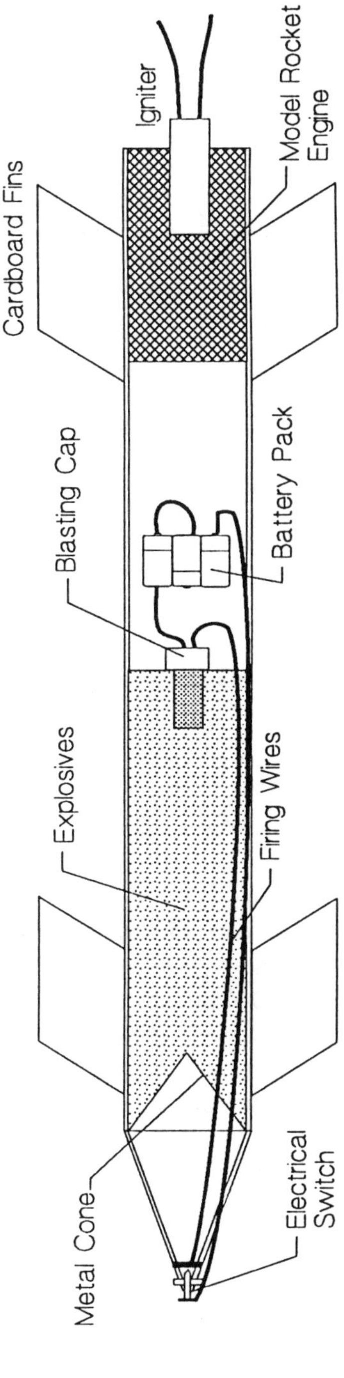

components described here can be purchased at hobby shops. Eight cardboard fins are glued onto the rocket body so it fits snugly into the launch tube. The fins help to stabilize the rocket in flight. Fold a cone from sheet metal and attach it to the front of the rocket after filling the rocket body with burster, flash powder, or high explosive. If the body of the rocket is packed with plastique or dynamite, place a booster explosive in the end near the nose. In the nose of the rocket, fasten a pistol or rifle cartridge, with the bullet removed, but with powder and a live primer, and glue over this a large cork with a hole drilled through it. Place a dulled nail in this hole so it can slide freely to contact the primer but is snug enough to stay in without falling out.

To fire the RPG, place the rocket into the launch tube and connect the igniter to the firing lines. Aim and hit the switch. The rocket motor drives the grenade from the tube. When it hits its target, the nail is driven into the primer of the cartridge, setting it off and detonating the booster charge.

The rocket should be packed with about a pound of high explosive.

The RPG can have an increased effect by using an electrically detonated shaped charge in the warhead.

Fill the body of the rocket with a pound and a half of high explosive, booster and blasting cap. The blasting cap must be at the rear of the explosive charge. Behind this, a battery pack is placed, with wires running to the front of the rocket.

Form a metal nose cone four inches long and attach it to the front of the rocket. This cone is hollow; its only purpose is to streamline the rocket in flight and to provide proper "stand-off" distance for the shaped charge. Fasten a small metal plate to the front of the nose cone and wire one of the blasting cap firing lines to this. Fasten a nail through a similar plate and wire it, separating the two with a styrofoam spacer. Impact on the target drives the nail through the styrofoam, detonating the explosive.

This rocket is effective against lightly armored vehicles and trucks. It is not capable of penetrating modern tank armor. Regular high explosive rockets are effective against buildings and bunkers.

All of these grenade weapons should be built carefully and each component should be tested separately to ensure safe construction and operation. Sand-filled rockets should be test-fired to familiarize the guerrilla with operation of the improvised RPG.

FIREARMS

The primary weapon of the guerrilla is, of course, the firearm. Weapons suitable for rural guerrilla ambushes are military assault rifles such as the M-16, AK-47, Galil or the G-3. These weapons are long-range and capable of single-shot accurate fire.

Urban guerillas would do well with small, easy-to-conceal submachine guns such as the Ingram MAC 10 or the 9mm Uzi. Assault rifles such as the M-16 or AK-47 are also acceptable. Sidearms should include the .38 revolver, the 9mm automatic or the .45 automatic. Small, concealable handguns such as the Colt Mustang or the Walther PPK are also ideal urban guerrilla arms.

Guerrillas, both rural and urban, should seek to use the same weapons as the regime's armed forces, allowing them to utilize captured weapons, ammo and parts. In the case of guerrillas fighting Soviet-supplied troops, the weapons of choice are the AK-47 and AK-74 assault rifles, the RPK light machine gun and the Makarov pistol. Resisters against US-supplied troops must be familiar with the M-16 rifle, SAW light machine gun and Colt .45 and Beretta 9mm pistols. American and West European police forces are usually armed with .38 revolvers or 9mm automatics, 12-gauge shotguns, and a few M-16/AR-15 rifles and Uzi carbines.

In some circumstances, an urban guerrilla unit may find itself facing hostile forces without a weapon. Even in these circumstances, serviceable firearms can be improvised from ordinary plumbing supplies.

Using commonly available materials, the guerrilla armorer is capable of turning out crude but workable submachine guns which fire on fully automatic. Rough plans for such a SMG are described here. All measurements are approximate, and much hand-fitting must be done during construction.

The most difficult part to manufacture is the magazine, and if it is at all possible a manufactured magazine should be obtained, with the rest of the gun fitted around it.

Small, light clips are ideal. If the gun is to be manufactured in .22 caliber, the clips from a Ruger Mini 14 or 10/22 will work well. The best caliber for SMGs is 9mm, and a magazine from an Uzi carbine works well for this design. Since the quality of the steel parts used in the gun is likely to be inferior, it is prudent to use .380 (9mm short) ammo instead of full-powered 9mm Parabellum. An Uzi magazine will accept and feed .380 auto cartridges as well as 9mm Parabellum.

The bolt of the SMG is made from a piece of steel bar measuring one inch in diameter and four inches long.

Clear a portion from the front of the bolt to clear the magazine lips, and grind a fixed firing pin at the front. A slot is also cut for an extractor and spring.

The receiver is made from one-inch inner diameter steel pipe, threaded at one end for a pipe cap. This must be screwed on tightly and welded into place. Cut the indicated holes in the receiver for the trigger/sear, the magazine port and the ejection port. Weld an ejector in place inside the receiver tube.

The barrel is a length of quarter inch inner diameter pipe with threads at one end. Bore a chamber just large enough for the cartridge to fit inside, then screw the barrel into a coupler which fits snugly inside the receiver. Screw this into place with a series of metal screws.

A trigger housing is bent from sheet metal and soldered into place, with the trigger and sear pinned inside. Screw a rear pistol grip into this and attach a wire stock to the pistol grip. Solder a magazine housing to the front over the magazine port, and add the magazine latch.

To assemble the weapon, screw the pipe cap onto the rear of the receiver. Slide the recoil spring inside and slide the bolt in on top of it. Screw the cocking handle into the bolt. Screw the barrel coupler into place at the front of the receiver and thread the barrel into position. Solder the magazine housing into place and the trigger housing. Pin the trigger and sear into position and screw the hand grip into place. Finish assembly by attaching the wire stock to the hand grip.

SUBMACHINE GUN COMPONENTS

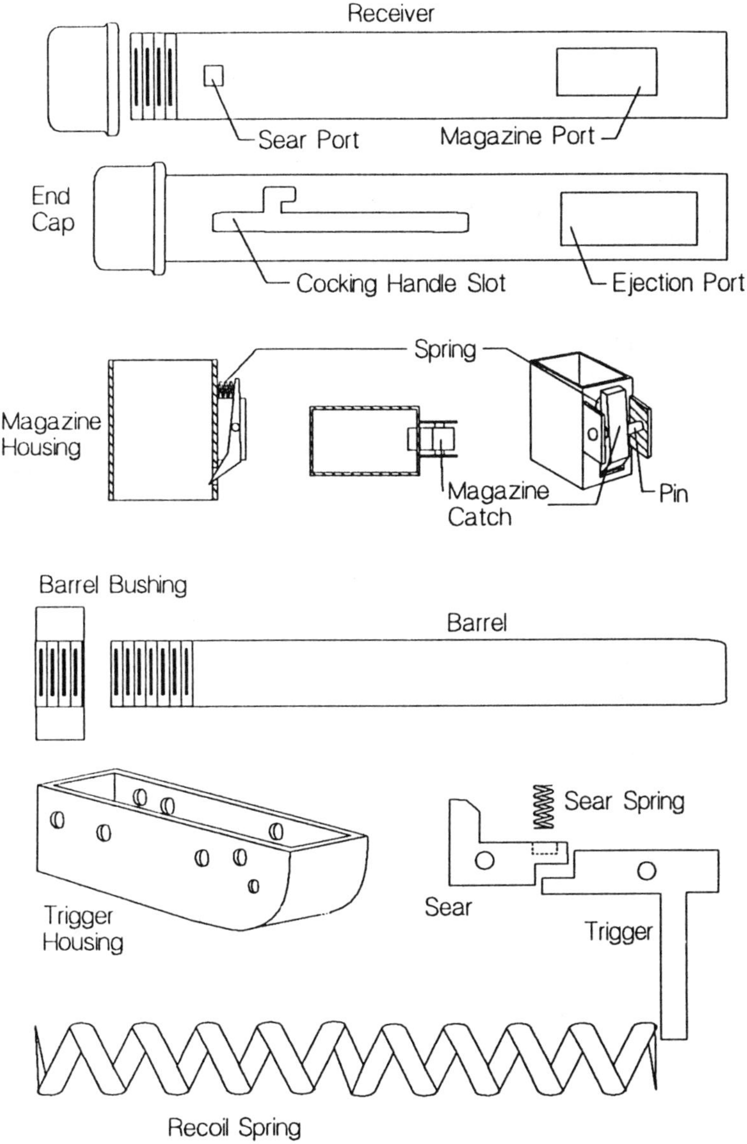

SUBMACHINE GUN COMPONENTS

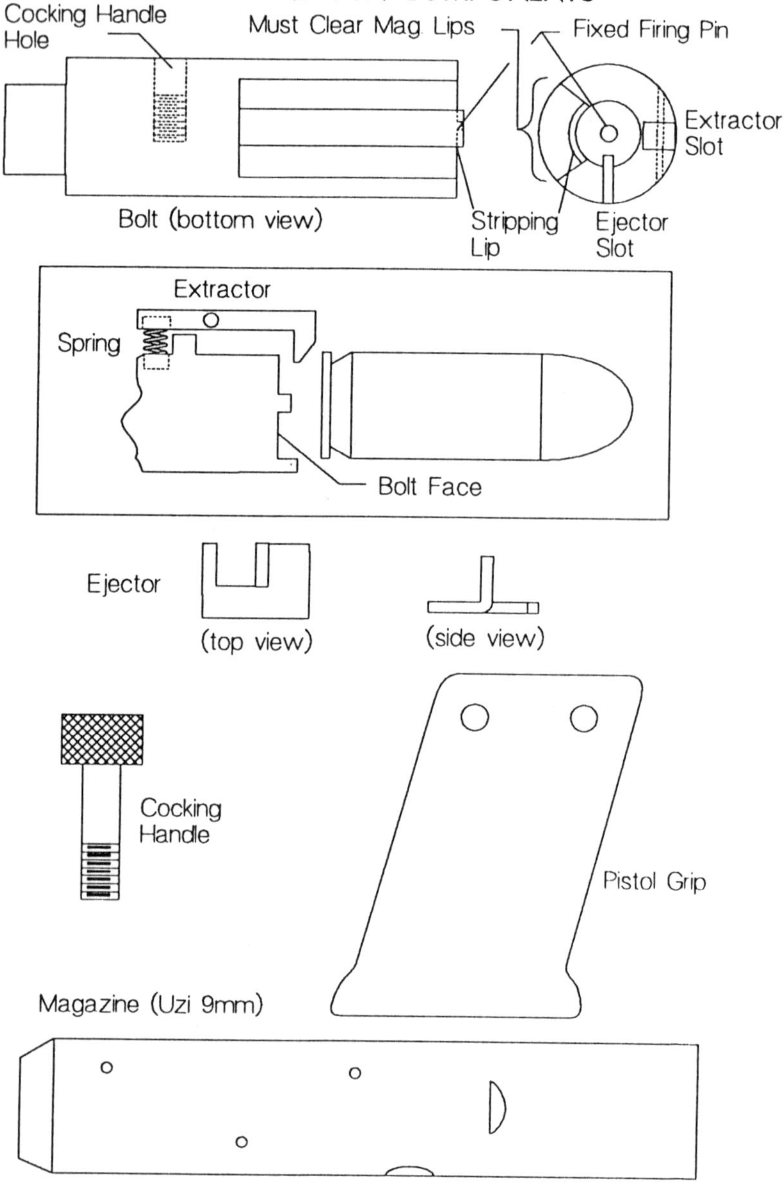

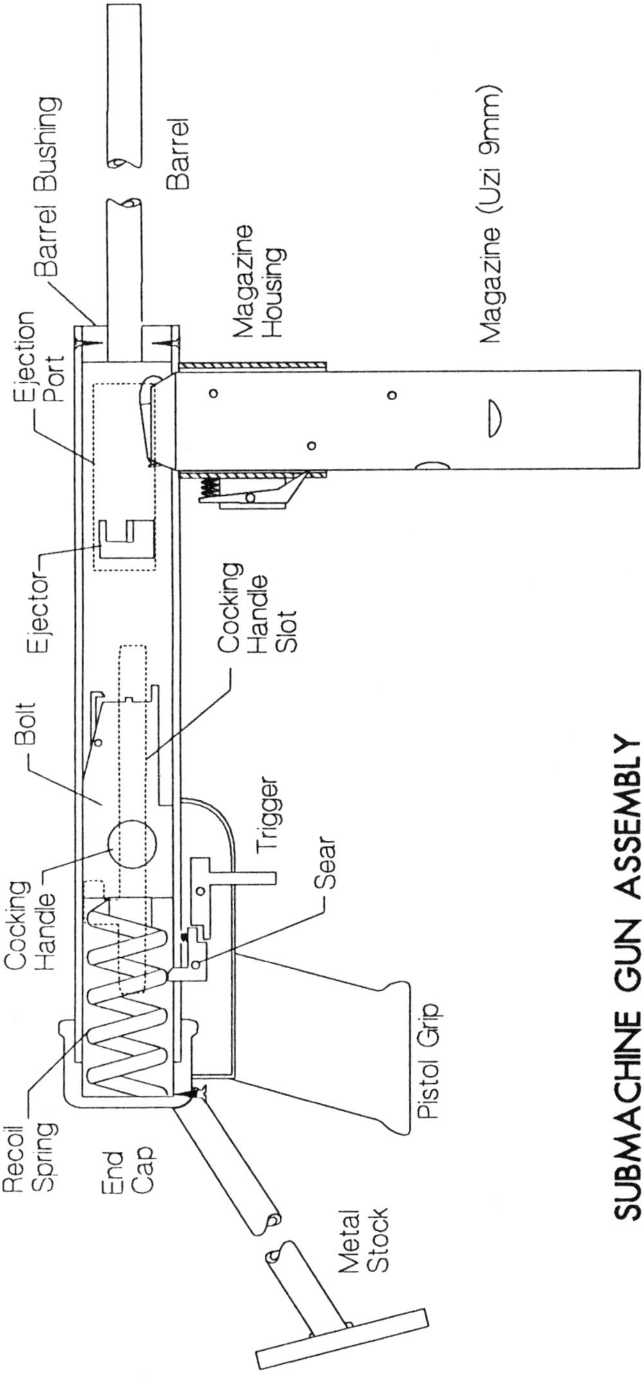

SUBMACHINE GUN ASSEMBLY

To fire, pull the bolt all the way back and pull the trigger. The weapon will fire until the trigger is released.

The urban guerrilla can also manufacture crude single-shot firearms for use in raids or ambushes. Single-shot pistols, such as the "zip" guns used by streets gangs, require the simplest materials but are lethal at short range. All that is needed are an old mousetrap, a piece of wood, a length of steel pipe and a pipe coupling or pipe cap.

The gun is assembled as shown in the diagram. The best springs come from old mousetraps, but in a pinch any strip of flexible metal will do. The pipe is tied firmly to the wooden stock with wire and tape. Steel pipe with a quarter inch bore will accept 9mm pistol ammunition. The pipe in this case is big enough for the bullet to enter, but too small for the cartridge case. If a pipe coupling is available, this is screwed onto the end of the barrel far enough so the primer of the chambered pistol cartridge projects slightly. If a pipe cap must be used instead, the pipe must be widened for a short distance to serve as a chamber for the round. The pipe cap has a hole drilled in it so a firing pin will reach the chambered primer. A nail or tack, dulled with a file and soldered to the mousetrap spring, serves as a firing pin.

To fire the gun, pull the spring back and hold it with the thumb. Aim and release the spring. The firing pin strikes the primer, discharging the bullet. To reload, unscrew the pipe plug fitted into the coupling or the pipe cap, and use a length of wooden dowel to push out the spent cartridge.

A serviceable single-shot shotgun is made in a similar manner. A larger, more powerful spring should be used, and the stock is longer to hold the barrel. A pipe with a 3/4 inch bore will hold and fire 12-gauge shotgun shells. The barrel should be between 13 and 15 inches long.

A single shot weapon using standard 7.62mm (.30 calibre) NATO or Soviet rifle ammunition can be made from common plumbing fixtures. The guerrilla will need two feet of 1/4 inch steel pipe, a 3/8 inch plumbing coupling and a pipe plug which will fit inside it, a 3/8 inch to 1/4 inch pipe reducer, a mousetrap, wire and a wooden stock. Use wire to securely bind the barrel to

SINGLE SHOT FIREARMS

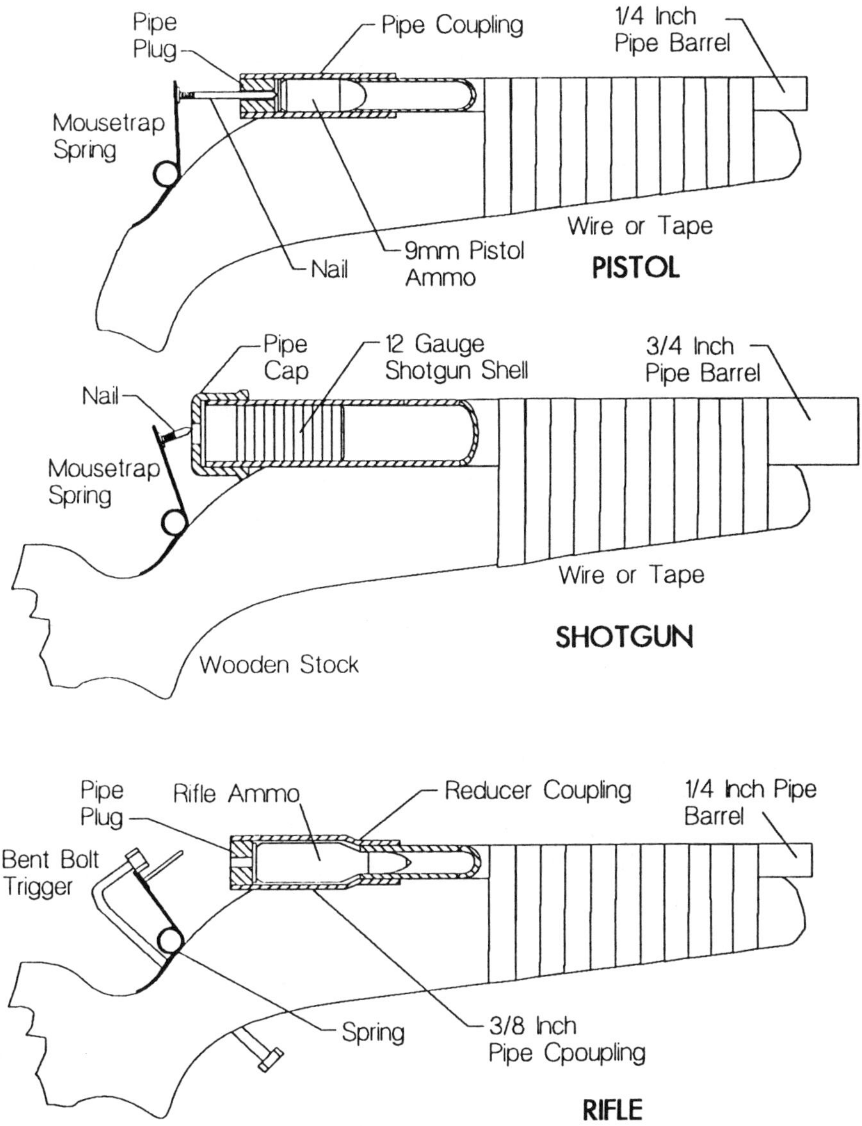

the stock. Thread the reducer onto the barrel and thread the coupling onto the reducer. Drill a hole through the pipe plug so the firing pin will reach through it to the primer of a chambered round. The spring is arranged as for the shotgun.

To use this weapon, load the rifle cartridge into the chamber. The bullet will fit into the barrel, while the wider powder chamber will stay in the coupling. Screw the pipe plug into place. Hold the spring back with the thumb, aim and release the spring.

All of these improvised weapons may be provided with crude triggers by adding a bent bolt or long nail. Drill a hole through the stock so that the head of the nail will catch the spring, with part of the nail projecting from the bottom of the stock. To fire, aim the weapon and push up on the bottom part of the nail, releasing the trigger.

In certain circumstances, the guerrilla may need to make and utilize these weapons when ammunition is scarce or in limited quantities. Therefore, the guerrilla must learn how to re-use expended ammunition cartridges with field expedient substitutes.

To prepare a used cartridge case for re-firing, use a long nail (blunted at the tip) and a hammer to push the fired primer cup and anvil from the bottom of the cartridge case. Hammer out the indentation in the pan caused by the firing pin. Cut the striking tips from several "strike anywhere" matches. Do not use the whole match head—just use the firing tips. Powder these and fill the primer pan with this powder. Tamp it tightly with a wooden stick. Place the anvil round-side-out in the primer pocket of the cartridge case and put the primer pan over it. Slowly squeeze the primer pan back into place using a vise.

Next, powder the match heads from which the tips were previously cut, and fill the cartridge case up to the neck with this powder. Tamp it in with a wooden stick.

To complete the cartridge, saw off a length of 1/4 inch steel bar equal to the length of a standard 7.62mm bullet, and force this into the neck of the cartridge case. If possible, round the tip of this improvised bullet on a grinder. The cartridge case is now ready to be fired again.

Pistol and shotgun ammunition may be re-used in a similar

manner. Shotgun shells should be filled partway with the match head powder, followed by a layer of tissue or paper, and then BBs or tiny ball bearings. Improvised shot may also be obtained from the split shot used by fishermen. Care must be taken that the weight of the improvised shell load is not greater than that of a normal shotgun shell. If it is, the pressure in the gun may build up too fast, and the gun may explode. If the shell is to be fired in an improvised weapon, it is a good idea to reduce the powder load slightly.

In situations where no manufactured ammunition of any type is available, the guerrilla must improvise firearms which do not use standard ammunition. Securely fasten a two-foot length of 1/4 inch or 3/8 inch steel pipe to a wooden stock. Drill a small hole through a pipe cap and screw this onto the rear of the barrel. Powder about 45 safety matches and pour this powder down the barrel. Next, ram a wad of tissues or paper towel down the barrel until it rests on the powdered match heads.

Cut the striking tips from three matches. Place one of the tips inside the hole which has been drilled through the pipe cap. Tape the other two tips over this hole.

To prepare the gun for firing, ram a metal object such as a piece of steel bar, a ball bearing, or anything else which will fit snugly into the barrel, so that it rests on the tampering wad over the powder.

To fire the weapon, pull the spring back, aim, and let go.

The firing pin crushes the match tips, igniting them and causing flame to pass through the pipe cap and into the barrel, where it ignites the propellant. Effective range of this device is less than fifty yards.

An improvised match-propelled pistol can be manufactured using an ordinary cigarette lighter as a trigger. Bind the pipe barrel to the stock and cap it with an undrilled pipe cap. Drill a tiny hole through the bottom of the barrel so it projects into the barrel just in front of the pipe cap. Ram a match powder charge, cotton tamping and bullet into the barrel. The cigarette lighter is taped to the stock so that its flame will project into the hole drilled in the barrel. This ignites the propellant and fires the gun.

IMPROVISED CARTRIDGELESS FIREARMS

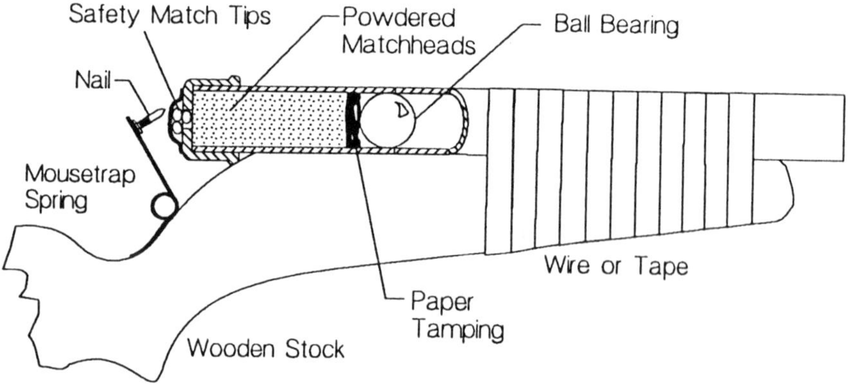

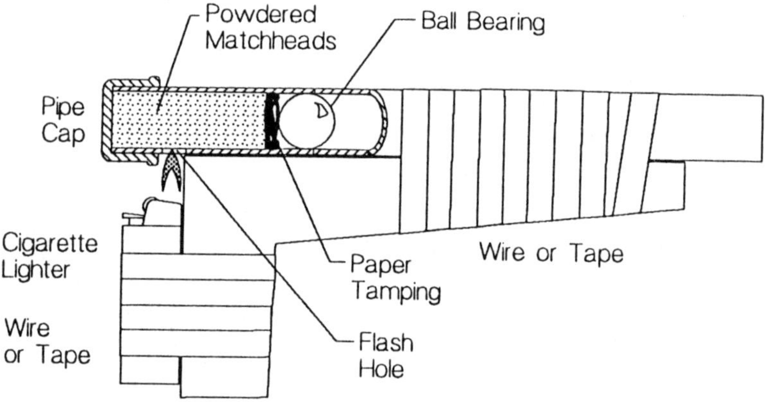

The improvised pistols and rifles described here can all be fitted with improvised silencers for use in ambushes or assassinations. Single-shot weapons are best suited to the use of silencers because they produce little mechanical noise, and few sources of escaping gases.

Weapons to be silenced should have threads at the muzzle end. The silencer is made using copper scouring pads, an empty soup or soda can, and a length of pipe of the same diameter as the improvised gun barrel. Drill a series of holes through the silencer pipe, about a quarter inch apart, in four rows. Drill a hole through the bottom of the can and fit it over the drilled pipe. Pack the can tightly with copper scouring pads. Cut a hole in a metal lid and solder it in place over the silencer body. Use a coupler to thread the completed silencer onto the barrel of the improvised firearm.

The silencer works by reducing the expansion of the gases and slowing them, cooling them and lessening the sound waves produced by the explosion. This same silencer can be modified to mount on any rifle or semi-automatic pistol.

MISCELLANEOUS WEAPONS

The garrote is an easily improvised weapon which is silent and lethal. It is useful for eliminating enemy sentries in a clandestine manner.

The garrote can be improvised with a length of piano wire. Two metal or wooden handles can be fitted by drilling holes through short lengths of metal or wooden bars, then tying knots in the wire to hold them in place.

Pursuing enemy vehicles can be slowed by the use of caltrops. These are tiny metal objects which are so arranged that one sharp point always sticks straight up, no matter how they land. They can be scattered in the road to slow pursuers, or they can be hidden under the tires of parked vehicles to complicate pursuit.

The simplest caltrops are made by welding three sharp metal spikes in a pyramid shape, and then welding an additional spike on top.

Caltrops can also be manufactured using quarter-inch steel bar in five-inch lengths. Two lengthwise cuts are made in the bar at

IMPROVISED SILENCER

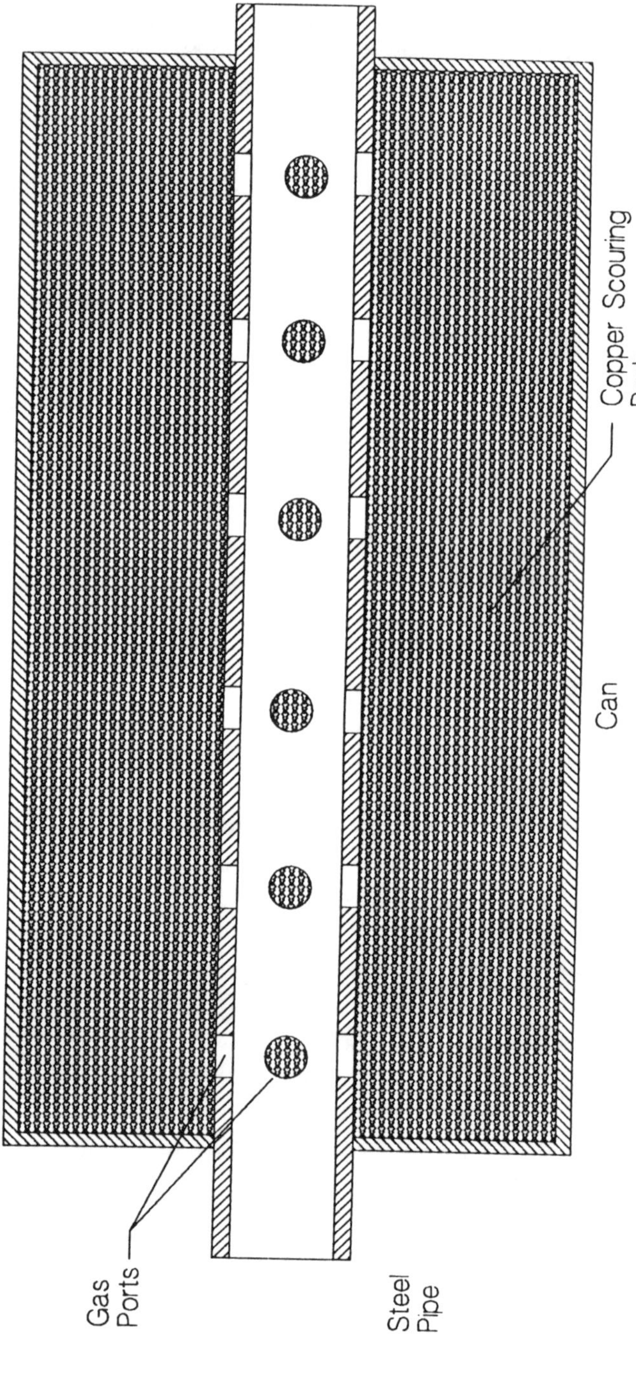

planes angled 90 degrees to each other, leaving the center section solid. These ends are then sharpened and bent outwards into the proper shape.

Simple caltrops are made from sheet metal. Cut a number of hourglass shapes about an inch and a half long, and bend them with a pair of pliers in a half twist, so that one corner rises straight into the air.

5

THE AMBUSH

The military actions of a guerrilla group must take the form of quick lightning attacks which strike unexpectedly and quickly disappear. Hit and run tactics allow the urban guerrillas to strike at the enemy without presenting a target for the enemy's counter-attack.

For this reason, the ambush is the primary method of engaging the enemy. In the initial stages of a guerrilla struggle, small enemy detachments are ambushed in order to obtain weapons, ammunition and supplies. As the guerrilla forces grow larger, more powerful enemy units can be successfully ambushed and destroyed.

The first step to be taken in planning a guerrilla action is to select an appropriate target. In many ways, the type of ambush to be laid has an effect on the tactics to be used.

The initial operations carried out by a guerrilla unit should be limited to simple and relatively risk-free actions, such as mining a stretch of road, ambushing a solitary enemy patrol or vehicle, or an assassination of an occupation official. These limited operations give the urban guerrillas a chance to learn to work with each other, and also to sharpen their operational skills. Success in these simple operations also gives a sense of confidence for future operations.

After the guerrilla team has gained operational experience and has captured suitable weapons, it must graduate to more complicated and dangerous tasks such as ambushing enemy supply convoys, troop columns and high-level authorities.

The urban ambush must be meticulously planned, keeping in mind the fundamental considerations of mobility, surprise, local superiority, and knowledge of the terrain. A good ambush site must provide cover and concealment for the ambushing forces while leaving little cover for enemy troops. It must provide a good all-around defense for the guerrillas and good fields of fire to saturate the ambush site.

The guerrilla's intelligence concerning the target must be complete. If the target of the ambush is an official of the regime or an individual vehicle, accurate information must be obtained concerning the schedule, route and any escorts or protection which may be encountered. Reconnaissance teams disguised as joggers, workers or innocent persons "going for a walk" must observe the target closely to determine its routine. Guerrillas planning an ambush on a supply convoy must know its route, time of departure, destination and speed of the convoy, as well as what supplies or materiel it is transporting, what weapons and armaments it carries for defense, and how many escorting vehicles it will have.

Once the target has been reconned, it is time to select an ambush team to assault it. The ambush team is divided into five-person firing teams, each with a nominated leader who is responsible to the ambush commander. These firing teams are grouped into four areas—rear support groups, close support groups, assault groups, and reserve groups.

The rear groups serve to isolate the area of action and to prevent enemy reinforcements from interfering with the ambush. They also serve as lookouts and early-warning sentries. After the ambush is completed, they function as rear guards to cover the withdrawal of the main guerrilla force. The rear support groups are armed with mines, grenades, at least one light machine gun, assault rifles or semi-automatic rifles.

The close support groups provide fire support for the ambush. They stop the enemy when the ambush is triggered and annihilate him with fire. The close support teams are armed with light machine guns, rifles, sub-machine guns, grenades and mines.

The assault groups are the actual fighting force of the ambush. Their task is to overwhelm the enemy after he has been weakened by fire from the close support teams. The assault groups are armed with rifles, submachine guns, shotguns and grenades. If the enemy attempts a counter-ambush assault, the guerrilla assault teams serve as a blocking force to trap the enemy in the ambush site.

The reserve groups are intended to reinforce any of the other groups if the need arises (this usually means reinforcing the assault groups if the enemy attempts to assault into the ambush.) It is armed with rifles, pistols or shotguns. After the ambush, the reserve teams function as stretcher bearers or porters who carry captured supplies and equipment.

In addition to these combat forces, the ambush team is accompanied by a detachment of medical personnel to treat casualties.

The number of firing teams needed in the ambush will vary with the circumstances. In narrow built-up areas, a smaller guerrilla team can cover the ambush site effectively. Ambushes of individual vehicles will require smaller guerrilla forces than ambushes on troops or supply convoys. The ambush forces should be just large enough to carry out their task, without being too large for easy command and control.

For security reasons, the urban guerrillas live dispersed in widely-separated safehouses. Once the ambush has been planned, the necessary fighters must be clandestinely contacted and assembled into firing teams. If possible, the firing teams should have an opportunity to fully rehearse the planned operation and practice for dealing with all possible circumstances. Since communication during the ambush will be difficult, the plan must be a simple and uncomplicated one, and each individual fighter must know his role and be able to perform it without guidance or direction.

On the day of the ambush, guerrilla fighters assemble at a pre-planned rendezvous point to receive weapons and final instructions. This RV should be some distance away from the ambush site, far enough to be outside the zone normally patrolled by the state security forces. The guerrillas leave all unnecessary equipment at the RV before proceeding to the ambush site.

A two or three person standing patrol should be left at the RV to provide rear security and to safeguard it against enemy forces. This team should be in radio contact with the ambush commander to warn if the RV is compromised before the fighters return.

When the ambush team is armed and assembled at the RV, it must occupy the ambush site and wait for the target to appear. This process must be absolutely clandestine to prevent informers or spies from spotting the ambush and giving warning. The ambush site is occupied slowly by two or three persons at a time.

The rear support groups move into position about half an hour before the raid. These teams occupy any potential routes which may be taken by an enemy column sent to reinforce the target. Highways, bridges, foot paths and railroad tracks should be watched by rear support teams, and ambushes laid for any approaching enemy forces. If an enemy force does approach, the rear support team's task is to hold them off long enough for the ambush team to complete its job and withdraw.

The rear support team must take care to stay well-hidden, in a position from which it can attack and stop enemy forces with rifle and machine gun fire, as well as grenades and Molotov cocktails.

Once the rear support teams have assumed their positions and established contact with the ambush commander, the close support teams infiltrate into their positions at the ambush site. These support teams should be close enough to the ambush site to provide effective and concentrated fire which will saturate the site. If close support teams are stationed along both sides of the ambush site, they should be at differing levels to avoid the danger of crossfire, i.e., the teams on one side of the road should be on the first floor of these buildings, while those on the other side should be on the second or third, out of the line of fire of the first team. In this way, concentrated fire can be delivered to all points of the ambush zone.

The ambush team's machine guns should be emplaced so as to seal off both ends of the ambush zone and to cover it completely with fire. If no machine guns are available, groups of riflemen must serve this purpose instead.

After the close support teams are in position, the assault group moves in. The assault teams must be in a protected position as close as possible to the ambush zone. If mines, roadblocks, etc., are to be used in the ambush, they are emplaced by the assault teams under cover from the close support guerrillas.

Once the site has been occupied and the ambush set, the guerrillas must wait until their target enters the ambush zone. This

POSITIONS OF REAR SUPPORT TEAMS

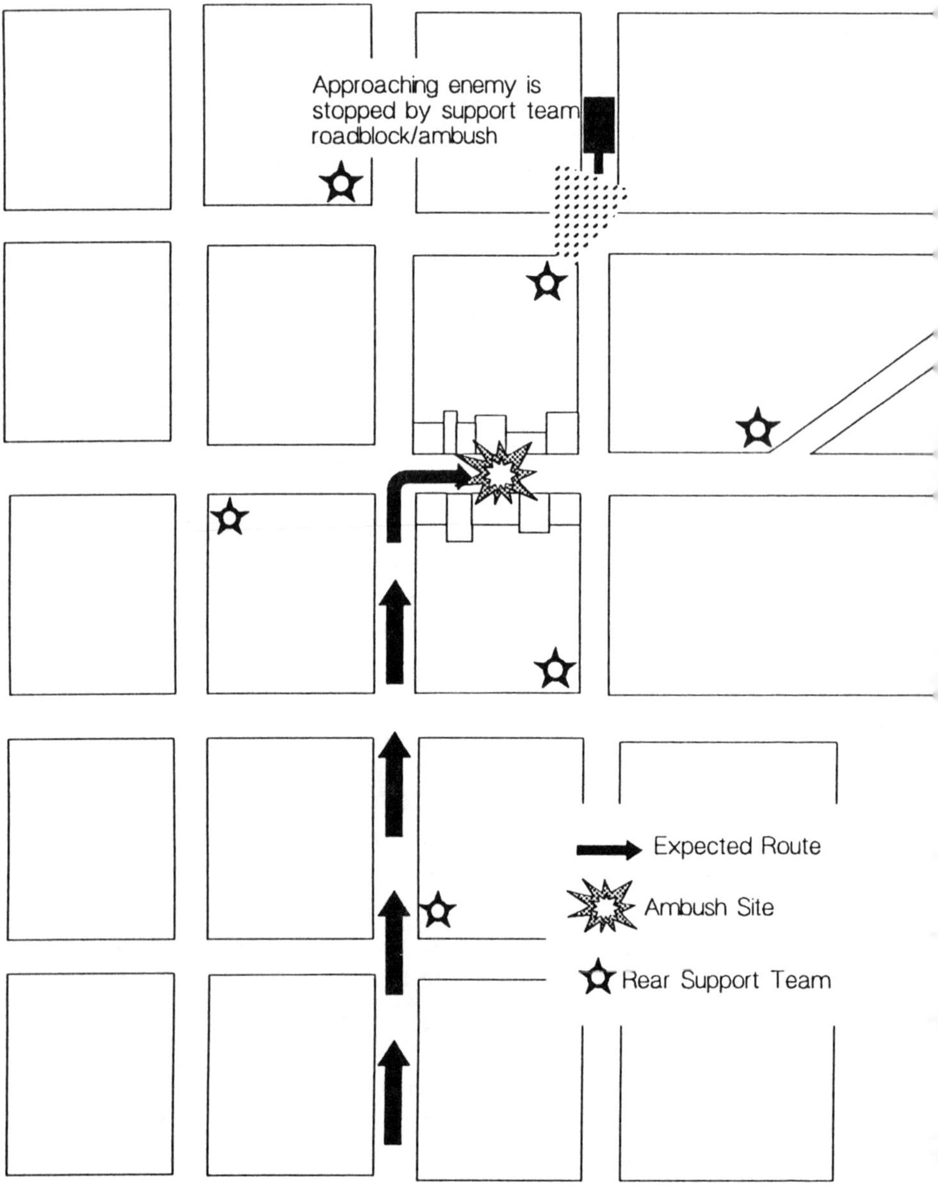

requires a tremendous amount of discipline. The guerrillas must remain absolutely invisible and silent to avoid giving away the ambush. Snipers must stay away from windows or doorways. Buildings and rooms which are occupied by the ambush forces must be kept secure, and any civilians who are found there must be held securely and silently until after the operation.

In many instances, exact intelligence information may be difficult to obtain and it may not be known what route the target may be taking. In these circumstances, it is necessary to emplace an ambush encompassing the entire area of possible approach. This is done most easily by setting up a wide net of rear support lookouts and choosing several possible ambush sites within this area, each manned by a separate ambush team. When the lookouts spot the target and determine its route, the appropriate ambush teams are alerted and, when the target enters an ambush zone, it is attacked.

This type of operation requires more careful planning than the simple ambush. Each team should be in radio contact with each other to coordinate their actions. Withdrawals should be planned to separate RV's and every possible problem must be anticipated and planned for.

The ambush may have as its goal any of several options. It may have the purpose of destroying the enemy detachment completely and capturing a supply of weapons, ammunition or other supplies. It may be designed to attack and loot a portion of a larger enemy column or convoy. It may have as its goal the capture of a prisoner for interrogation or intelligence purposes. Or it may simply have the goal of hitting the enemy to cause maximum casualties before hastily withdrawing.

When the rear support teams warn of the approach of the target, the ambush team must ready its weapons and prepare to spring the ambush. There must be some definite signal which marks the beginning of the ambush action.

This may be a shot fired by the commander, the detonation of a mine or hand grenade, or a burst of fire from the machine gun. It is vital that no actions be taken until this signal is given, since premature firing will alert the enemy to the existence of the

POSITIONS OF CLOSE SUPPORT TEAMS

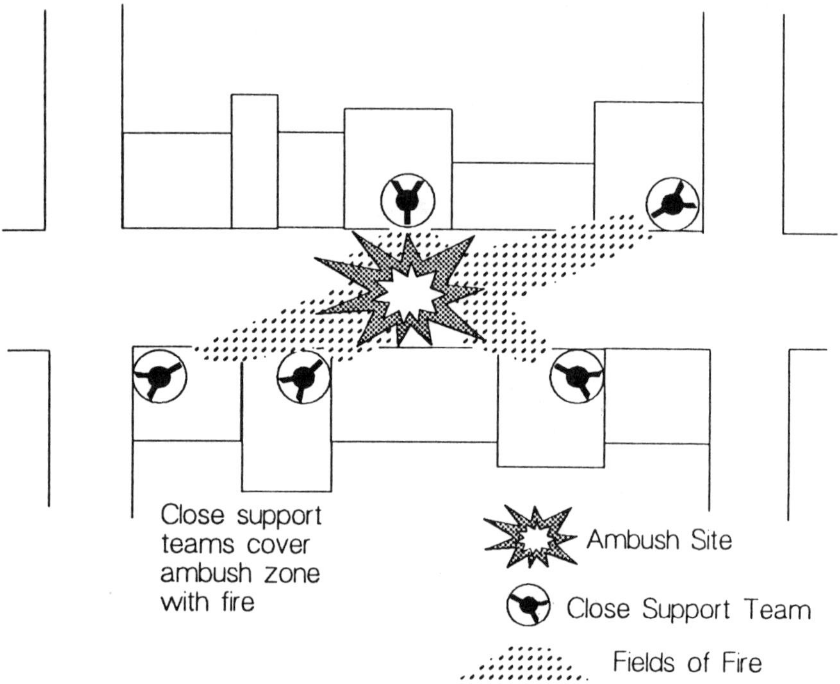

POSITIONS OF ASSAULT TEAMS

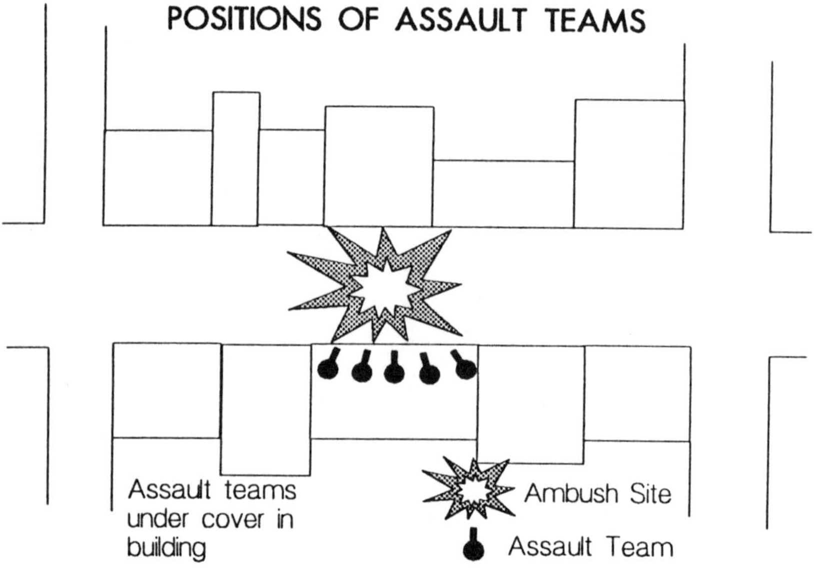

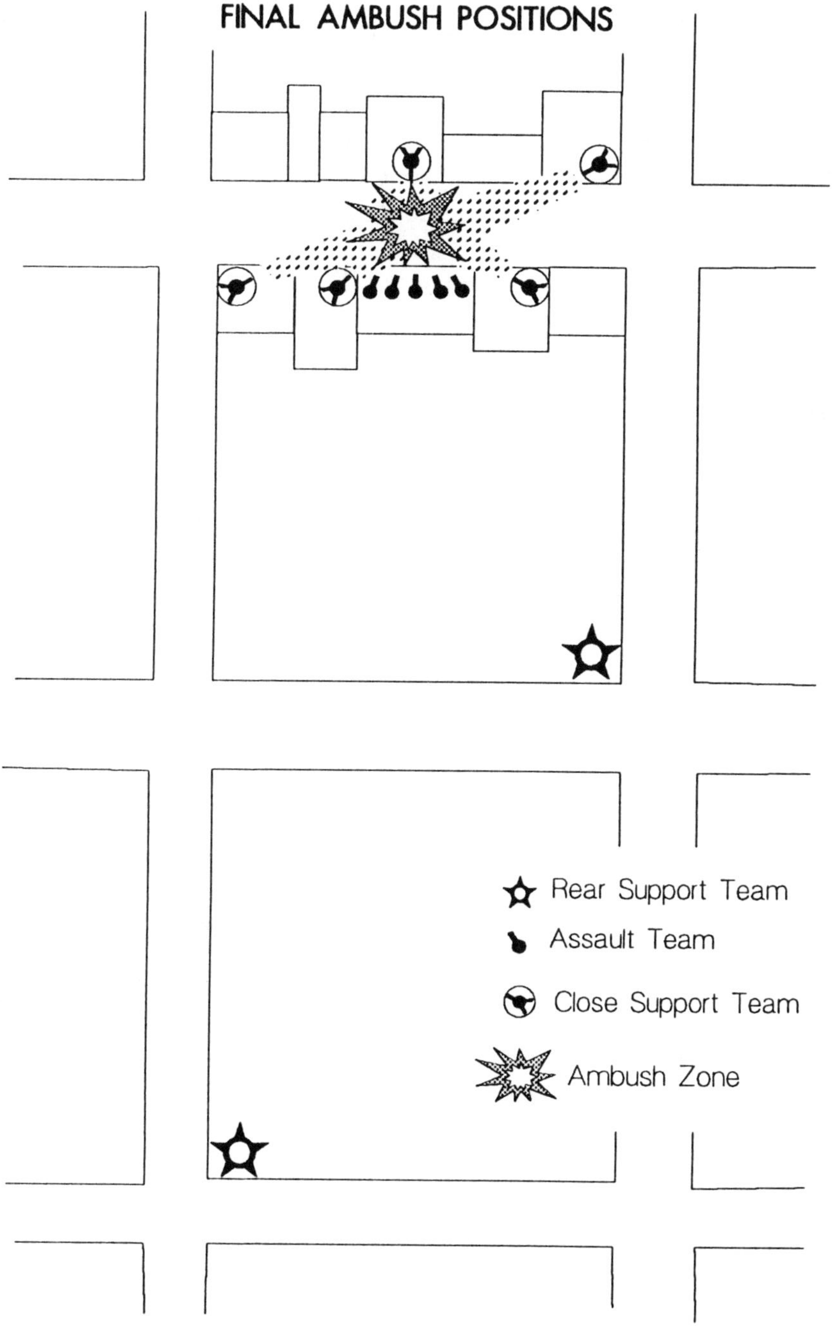
FINAL AMBUSH POSITIONS
Rear Support Team
Assault Team
Close Support Team
Ambush Zone

ambush before he is completely within the ambush zone. If, however, anyone in the enemy force acts as though he has seen the ambush, fire must be opened immediately.

Once the ambush is sprung, the ambush teams open fire, filling the ambush zone with intense fire which decimates the enemy. Fire must be distributed throughout the enemy column, with one firing team attacking the head of the column, one attacking the rear and another firing upon the middle. The object of the ambush is to produce the maximum amount of fire in the minimum amount of time to cover the ambush zone completely.

After the initial burst of fire, surviving enemy forces will scatter and attempt to seek cover in buildings, behind vehicles, etc. These survivors must be picked off by the assault teams waiting in the nearby buildings.

In order to prevent the arrival of enemy reinforcements to the scene, the duration of the guerrilla action must be as short as possible. Therefore, provisions must be made to signal an end to the ambush and the beginning of the withdrawal. If the goal of the raid is simply to hit and run, the signal for withdrawal marks the end of the ambush action. The withdrawal of the guerrilla force should not be immediate; instead, the guerrillas exfiltrate in small groups of two or three, covered by fire from the rest of the teams. This covering fire should prevent the pursuit of the guerrillas by the enemy.

The ambush team retreats to the RV in small groups and assembles there. In retreating, guerrillas follow pre-planned routes which are covered by the rear support teams. The rear support groups act as rear guards, to ambush and halt any pursuing enemy forces.

If, on the other hand, the purpose of the ambush is to obtain captured supplies or prisoners, a follow-up action is needed after the target has been decimated by fire. This follow-up is the responsibility of the assault teams.

If a follow-up assault is needed, fire from the ambush team must cease in an instant after the target has been saturated with fire. When the ambush is sprung, the enemy will be confused and disoriented by the noise, the surprise and the casualties. If the

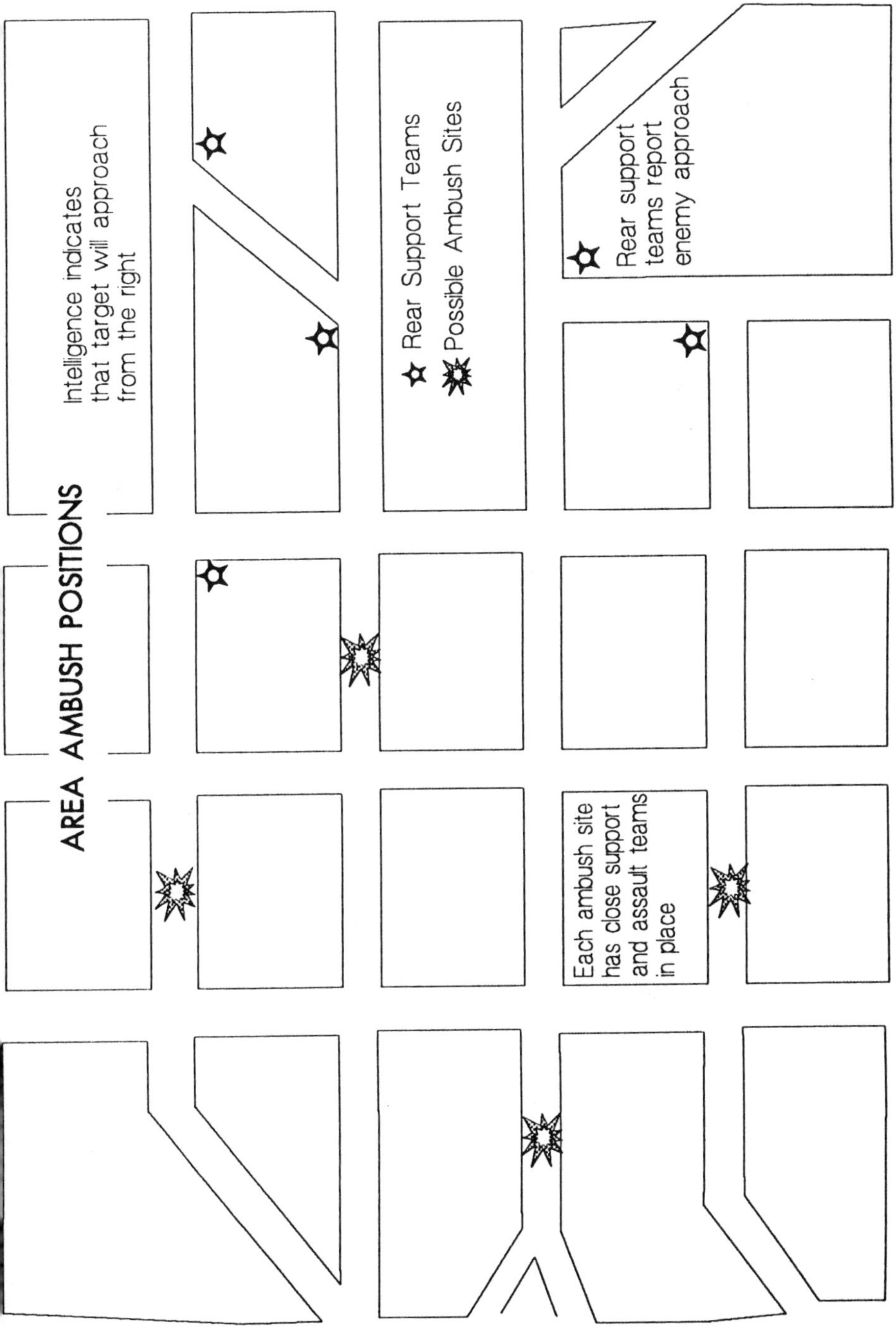
AREA AMBUSH POSITIONS
Intelligence indicates that target will approach from the right
Rear Support Teams
Possible Ambush Sites
Rear support teams report enemy approach
Each ambush site has close support and assault teams in place

enemy commander is killed or wounded in the initial burst of fire, it may take some moments for the enemy to establish discipline. The follow-up assault must take place during this moment of confusion.

The instant that fire from the ambush teams cease, the assault teams move into the ambush zone to overwhelm and eliminate any surviving resistance. Once resistance has been eliminated, the assault troops begin to gather enemy weapons and supplies, under the covering fire of the close support elements. At the same time, medical teams move in to deal with casualties and transport them back to the RV. In some raids, unarmed guerrillas may follow the assault teams to gather and transport weapons and supplies.

After available supplies have been gathered, the assault and transport teams withdraw in small groups to the RV. The close support teams withdraw next, followed at last by the rear support teams.

At the RV, the captured equipment is taken by guerrilla sympathizers to the place where it will be stored. The guerrilla fighters withdraw individually to their safehouses.

The guerrillas must make allowances for variations in this basic ambush plan. A signal must be given to end the ambush and withdraw without action if the enemy approaches in force stronger than expected or if the enemy fails to appear at all. A signal for withdrawal must also be agreed upon if the ambush begins to turn against the guerrillas.

Night ambushes may be carried out by the guerrillas in much the same manner as daytime actions. Some provision for illuminating the ambush zone must be made. If possible, an area that is illuminated by street lights is ideal. If no light is available, the guerrillas must use flares, car headlights, or some other illumination at the moment the ambush is sprung.

Also, since accurate shooting is more difficult at night, the guerrillas must use automatic weapons and shotguns, which can cover large areas at a time and thus increase the firepower of the ambush. Increased use of grenades will also increase the firepower of the guerrilla team.

Unless the ambush is planned specifically for capturing an

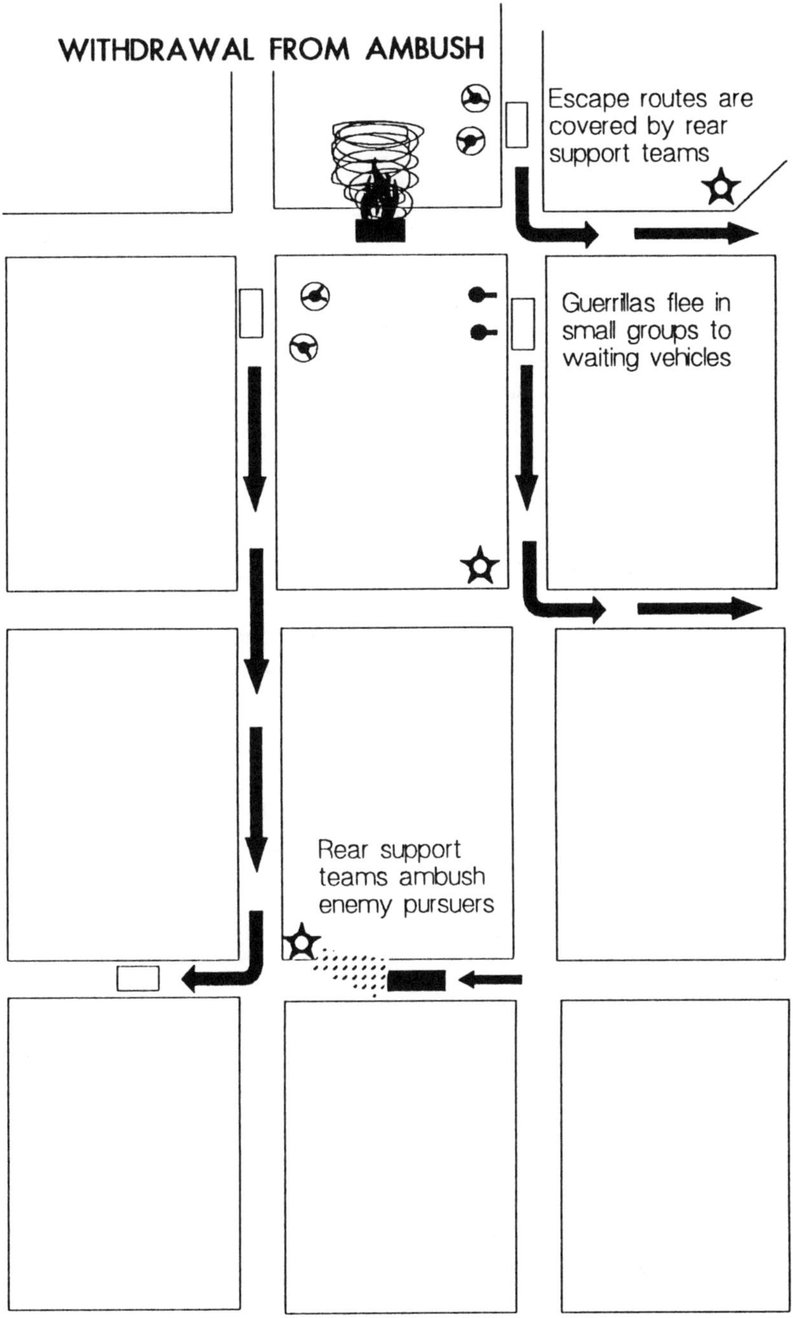
WITHDRAWAL FROM AMBUSH
Escape routes are covered by rear support teams
Guerrillas flee in small groups to waiting vehicles
Rear support teams ambush enemy pursuers

enemy prisoner, guerrilla units should make it a practice to take no prisoners in their operations. Prisoners require a secure area for holding them captive, and guerrilla warfare has no place for secure areas.

Captured troops should be stripped of all useful possessions—weapons, ammo, canteens, boots—and then released. Not only will this be a method of obtaining supplies for the guerrilla forces, but the released soldier will talk to his comrades, who, when faced with combat, will be more inclined to surrender to temporary captivity rather than fighting to the death. Thus the enemy's fighting spirit and morale are undermined.

The raids carried out by urban guerrillas are, in essence, nothing more than an ambush on a stationary target. Much of the planning for an urban raid will duplicate that for the ambush.

The type of target to be raided depends on the objective of the guerrilla action. In the initial stages of guerrilla warfare, the objectives are to obtain supplies, weapons and money for the resistance network. Weapons and ammunition are obtained by raids on arsenals, ammo dumps, small army posts or civilian police stations, or by ambushing enemy troops and supply trucks. Medical supplies can be procured by raids on hospitals, clinics and doctor offices. Cash can be expropriated by raids on banks, enemy payroll offices, or rental offices.

A large portion of guerrilla actions will have as their goal the destruction of enemy materiel or personnel. These actions take the form of raids on enemy communications centers, power stations, railroad stations and tracks, bridges and highways, fuel depots, outposts and security stations, factories and industrial centers, and government offices and officials.

On some occasions, guerrillas may wish to seize and occupy a target for some specific purpose. Radio or TV stations, for instance, may be temporarily occupied in order to broadcast messages to the population.

Once the target of the raid has been selected, observation and recon teams must study it to learn as much about its routine as possible. Some targets, such as radio and TV stations, are unlikely to be heavily guarded. Observation of these targets will simply determine the best time to move in and seize it.

Other targets, such as arsenals or fuel depots, are likely to be guarded by heavily armed security detachments. Among the information which must be gathered before a raid on such an installation is the location of guards and guardhouses, the guard routes and the times the guards are changed. The location of searchlights, machine guns or other heavy weapons, and other parts of the target's defenses must be noted. In the most high-security positions, the locations of heavy weapons will vary from day to night, as will sentry and guard positions.

Defensive perimeters will use a system of mutually supporting positions, listening posts and observation posts. If available, surveillance devices such as motion detectors, infrared sensors, etc., may be positioned. Wire, obstacles, patrol dogs and silent alarms may be used to prevent intruders.

Most military units use one squad of soldiers on patrol outside the perimeter (to watch for stand-off weapons such as grenade launchers or rockets) and one squad inside the perimeter. One machine gun will be at the main gate, and another will be located in a high tower or observation post. Riflemen patrol the perimeter constantly.

Targets in built-up industrial areas (such as police stations) do not have the space for an elaborate defense perimeter. They are likely to be protected by a roadblock or checkpoint at its approaches, armed guards at the entrances, and armed personnel inside.

The composition of the forces in a raiding party will differ somewhat from those used in ambushes. The rear support teams serve as lookouts and rear guards for withdrawal. They are armed with light machine guns, grenades, rifles or Molotov cocktails.

Close support teams provide covering fire for the area of operations. They are armed with submachine guns, shotguns, grenades and rifles.

The assault teams are the actual fighting force of the raid.

It is their job to eliminate enemy sentries or guards and to enter the objective. Once inside, they set up an all-around defense to allow the raid to be carried out.

REAR AND CLOSE SUPPORT TEAMS

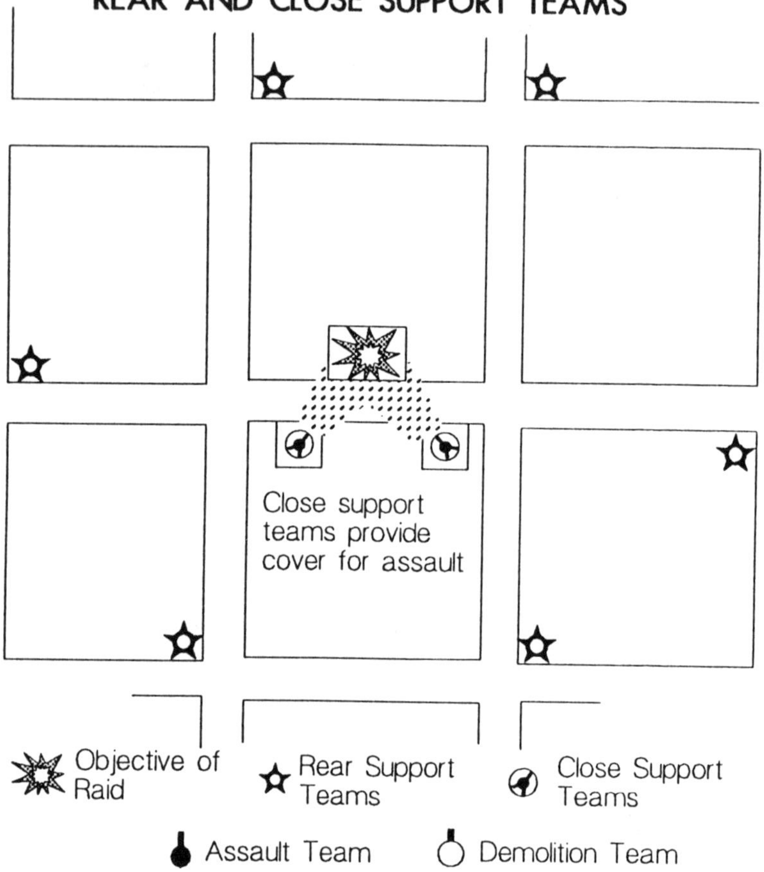

ASSAULT AND DEMOLITION TEAMS

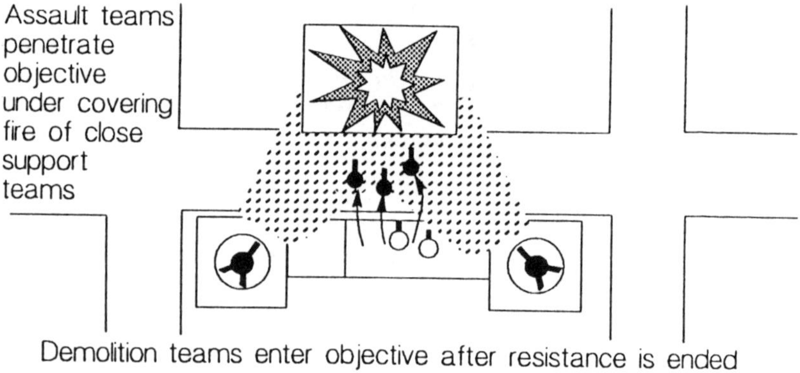

Demolition teams enter objective after resistance is ended

The demolition teams are the actual sabotage force of the raid. They are armed with pistols and submachine guns, but their primary role is to plant the explosives, demolition charges and sabotage devices which accomplish the purpose of the raid. They enter only after enemy resistance has ended, and are protected by the assault teams while they carry out their operations.

The guerrilla teams move into the area of the raid in much the same manner as they would occupy an ambush site. The rear support teams establish positions first, followed by the close support teams. The assault teams gather in position for attacking the target.

At the signal for attack, the assault teams move in to eliminate enemy resistance, supported by fire from the close support teams. Once resistance has ended, the demolition teams move in to plant explosive and sabotage the target.

In raids on supply bases and ammo dumps, the guerrillas load themselves with as much needed supplies as they can transport, then place sabotage charges to deny the enemy the use of the remaining supplies. Other targets, such as power stations, train stations, airfields, fuel depots and factories, are systematically destroyed by the demolition teams. In all raids, enemy vehicles and communications are destroyed or expropriated to hinder pursuit.

Attacks which aim at destroying a communications center should focus on the power supply and the control panel. The controls can be destroyed with a number of hand grenades or a half-pound demolition charge detonated on it. Destroy the transmitting tower with a number of explosive charges.

Power stations and electrical relays are disabled by destroying the transformers. These are made of thick metal plate, but can be pierced with an RPG, a shaped charge, or a demolition charge of plastique or four pounds of dynamite.

Railroad tracks can be cut with dynamite or plastic explosive charges. Guerrillas should blast away at least two feet of track. Locomotives are destroyed by explosive charges on the axles, and train stations are destroyed by blowing up the switching mechanisms.

WITHDRAWAL FROM RAID

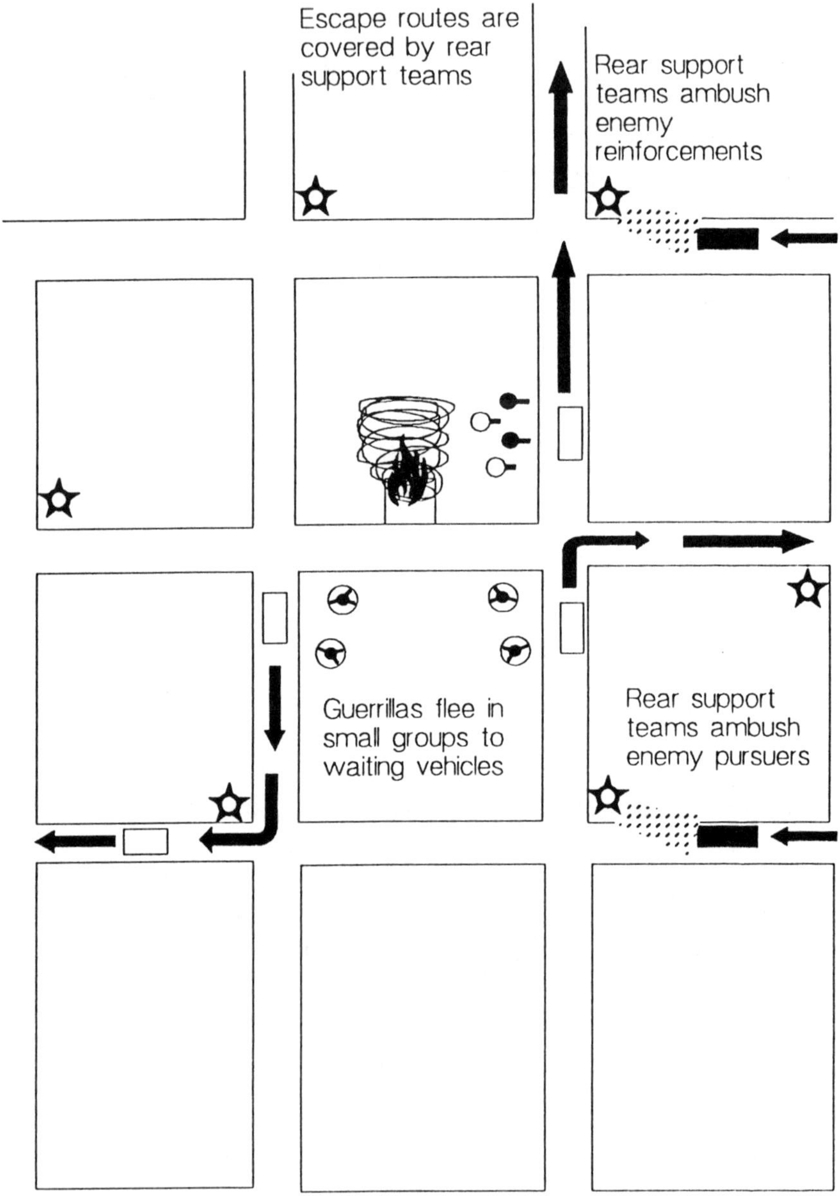

Fuel tanks and storage containers can be destroyed by detonating a plastique charge directly on them. This will ignite the fuel and cause a destructive fire and explosion.

When attacking an airfield, the primary targets are the aircraft themselves. Enemy aircraft can be quickly disabled by setting off a demolition charge on the nosewheel strut of each plane. For more thorough destruction, place a charge inside the jet engine intake. If the control tower can be penetrated, the radar equipment, radios and display terminals should be destroyed with explosives.

Helicopter gunships are one of the guerrilla's deadliest enemies and should be destroyed whenever found. Explosive charges in the cockpit will disable the chopper.

Factory machinery, or heavy equipment in a power plant, can be blown apart with heavy explosive charges, or can be skillfully sabotaged using Thermite grenades to weld precision parts together or to burn holes in heavy equipment. Any supply dump, ammo dump or fuel storage area is vulnerable to incendiary attack.

Once the demolition charges have been set and the purpose of the raid has been accomplished, the guerrilla force must quickly withdraw. The demolition and assault teams withdraw first, covered by the close support groups. Withdrawal routes are covered by the rear support groups, who intercept and ambush any pursuers.

It will become necessary, in some instances, to carry out guerrilla ambushes and raids in the rural areas to cut enemy supply and communication lines and to hinder the movement of enemy troops from one area to another. While the guerrillas must develop and perfect the principles of urban operations, they must not neglect the rural portion of the struggle.

Rural ambushes on supply trucks or small troop detachments are not very different from those carried out in urban areas.

The ambush site is selected and occupied by the rear support teams, the close support teams, the assault groups and the reserve forces. When the convoy enters the ambush zone, it is halted by the explosion of claymore mines or grenades, and enemy resistance is annihilated by fire and assault. The ambush party then withdraws before enemy reinforcements can arrive.

Once the guerrillas have carried out several successful ambushes against enemy supply trucks, the enemy will begin to move only in larger armed convoys, protected by APCs or light tanks. Since the guerrilla force is not strong enough to encircle and destroy the whole column, it must focus its firepower on one small part of it. The most effective way is to use a sharp bend in the road to isolate the lead truck or trucks and then attack them.

The mines and machines gun are placed at a point where the road bends sharply. The ambush party allows the first two or three vehicles to round the corner before triggering the roadblock, halting the lead vehicle.

At the same time, RPGs or more mines are detonated to prevent the rest of the column from rounding the bend. Machine gun and small arms fire prevents enemy troops from approaching the roadblock and the isolated vehicles, while assault teams attack the lead vehicles and loot them.

After the ambush has been completed, the guerrillas gather up supplies, wounded fighters and captured weapons and retreat in small groups. Using fire and maneuver to deter pursuit, the guerrillas exfiltrate in small groups to the RV, where they reassemble and move out of the area.

AMBUSHING A SUPPLY TRUCK

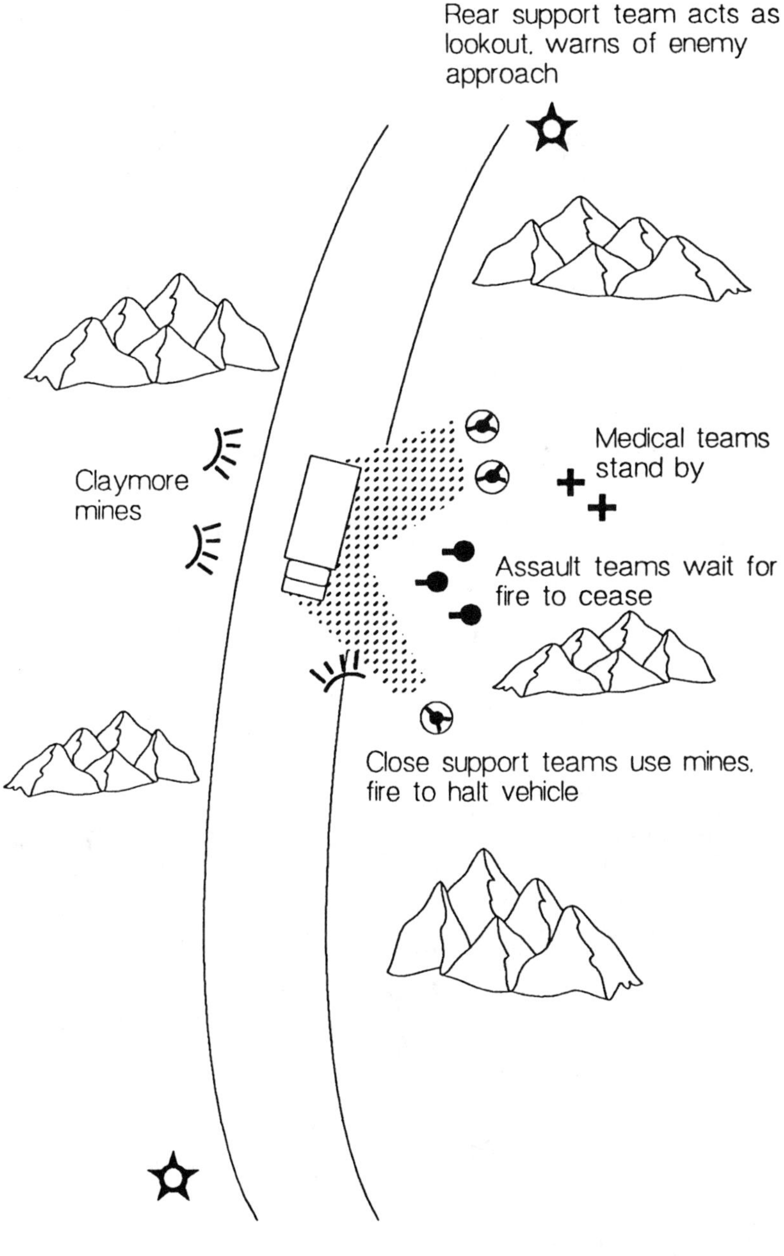

AMBUSHING A LARGE COLUMN

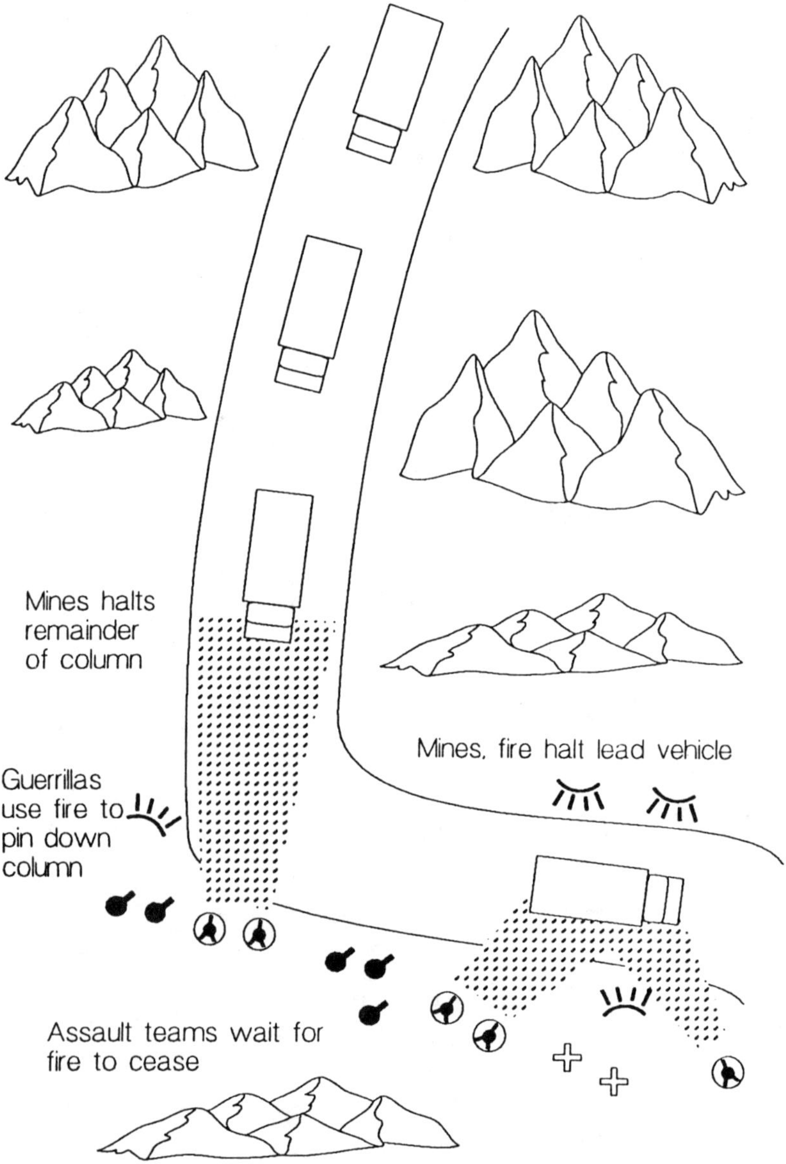

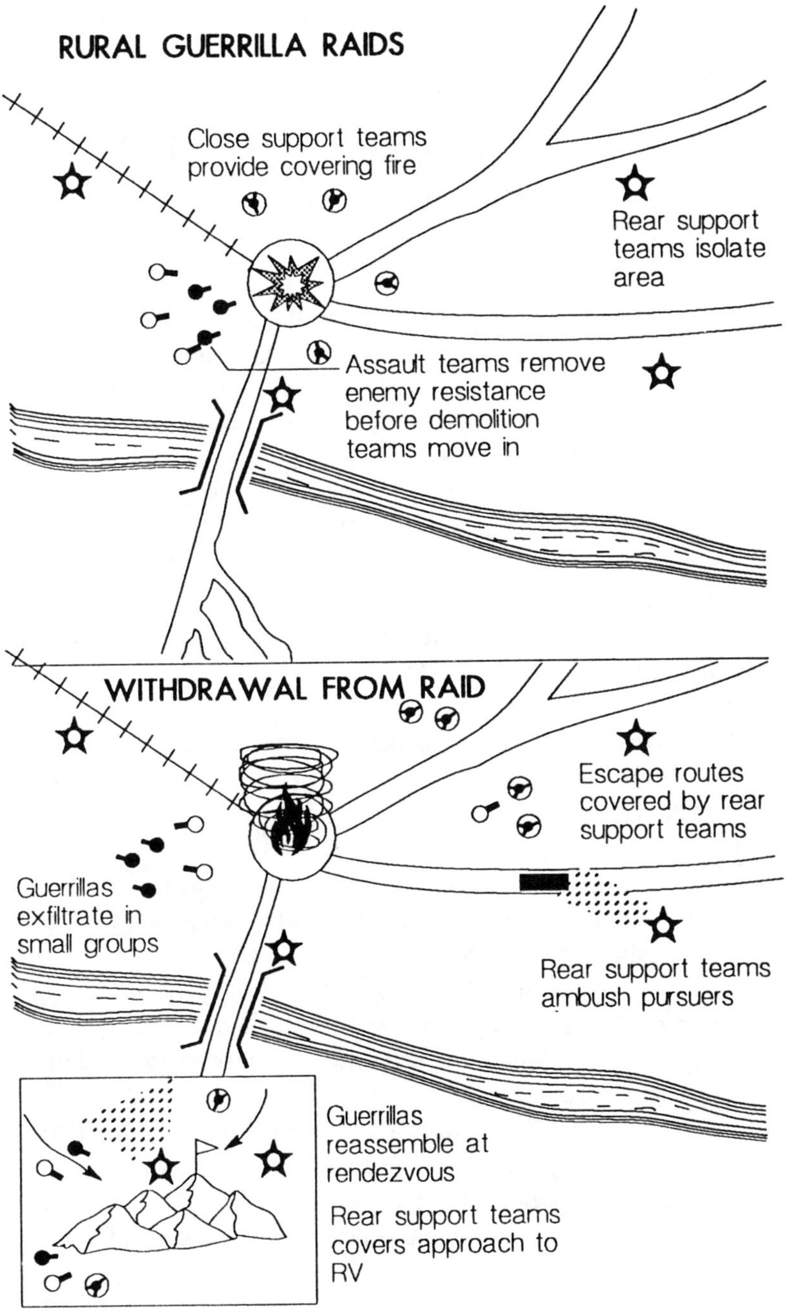
RURAL GUERRILLA RAIDS
Close support teams provide covering fire
Rear support teams isolate area
Assault teams remove enemy resistance before demolition teams move in
WITHDRAWAL FROM RAID
Escape routes covered by rear support teams
Guerrillas exfiltrate in small groups
Rear support teams ambush pursuers
Guerrillas reassemble at rendezvous
Rear support teams covers approach to RV

6

COUNTER-INSURGENCY OPERATIONS

The guerrilla network must be aware at all times that the state security forces of the dictatorship or the occupying power will be attempting to seek out the insurgents and destroy them. Urban resistance fighters must therefore be familiar with the counter-insurgency methods of the enemy and the ways to counter them.

Invading forces will be ruthless with irregular units and guerrillas. Captured resistance fighters will be shot without trial, usually in a public manner as a warning to the population. The invaders may select hostages from the people and execute them in retaliation for guerrilla raids and ambushes, hoping by this to frighten the population into betraying the guerrillas.

The guerrillas must not allow these actions to stop their operations. The invaders cannot be expelled and the dictatorship cannot be overthrown by giving in to its threats. Only armed action will topple the regime and end the oppression. While the guerrillas must do what they can to stop the execution of civilian hostages, they cannot stop their struggle because of these enemy actions.

The primary response of a military dictator to a guerrilla insurgency is military action. State security and armed forces will be directed to find, infiltrate and crush the resistance network.

Infiltrators and informers are the most dangerous enemies of a guerrilla unit. Spies can lay bare the entire operation of a resistance network, thus depriving the fighters of the secrecy and surprise necessary for their survival. For this reason, the counter-intelligence arm of the guerrillas must be strong and vigilant.

Steps must be taken to prevent the infiltration of the guerrilla network by spies and informers. Enemy agents must be ferreted out by strict security precautions. New guerrilla recruits must be carefully screened to filter out enemy infiltrators. Contact with civilians should be carefully monitored and limited. During periods of operations, all guerrilla contact with civilians must be suspended, lest information be inadvertently passed on to the enemy forces. All information about the guerrilla network must be made available only on a "need to know" basis.

Since the majority of guerrilla operations consist of raids and ambushes, the enemy will devote a considerable amount of energy towards counter-ambush tactics and training. The guerrilla teams must be prepared for these counter-ambush operations.

The most basic counter-ambush action takes the form of an assault by enemy troops into the ambush site, as soon as control has been re-established. The ambushed enemy troops will seek to escape the ambush zone as rapidly as possible by retreating into cover—doorways, behind walls, behind parked vehicles, and so on. Once they have escaped the ambush zone, they will begin to systematically attack, using fire and maneuver techniques, into the positions where the guerrillas have established themselves. By thus drawing the guerrillas into an open firefight, they hope to overwhelm the ambushers with their superior firepower.

The guerrilla must take steps to prevent this strategy from being successful. The fire of the close support teams into the ambush zone must be rapid, concentrated and must cover the ambush zone completely, to prevent enemy troops from escaping the field of fire and assembling for a counter-assault. Enemy troops which are seen fleeing towards cover must be picked off by the waiting guerrilla assault teams. If an enemy force does succeed in establishing itself and beginning a counter-assault, it must be met and countered by the assault teams acting as a blocking force. If this enemy is small and weak, it must be overcome by the assault teams and annihilated. If the enemy force is too great to be overwhelmed immediately, the guerrillas must not waste time in an involved firefight with it, but must give the signal for withdrawal to begin, with the guerilla assault teams acting as a rear guard to hold the enemy forces back until escape is complete. In most cases, the existence of an enemy counter-assault force will prevent the

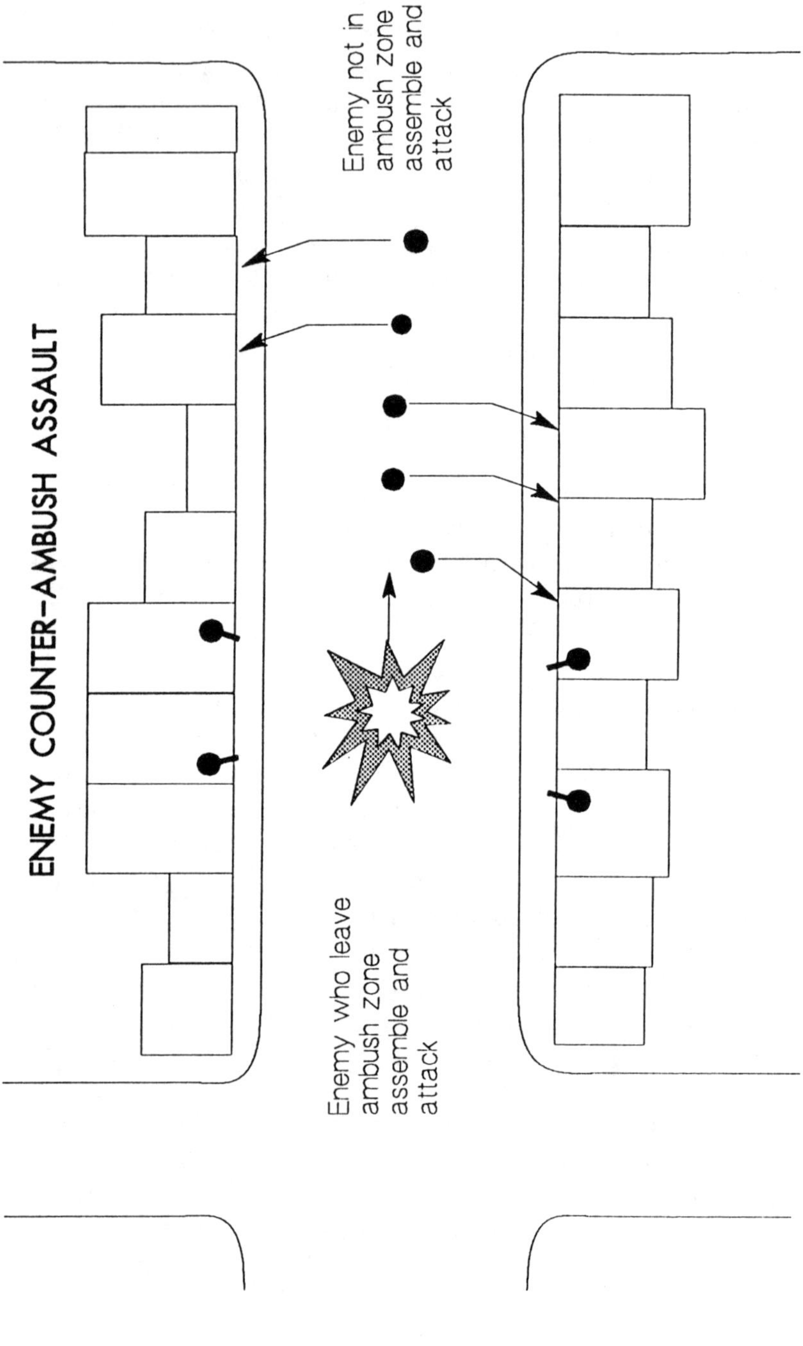
ENEMY COUNTER-AMBUSH ASSAULT
Enemy not in ambush zone assemble and attack
Enemy who leave ambush zone assemble and attack

objectives of the ambush from being carried out—the guerrillas must therefore prevent the enemy from leaving the ambush zone and establishing such a counter-force. This is best accomplished by carefully planning the ambush so that maximum firepower is concentrated on the ambush zone, instantly annihilating the enemy force.

Once the guerrillas have successfully ambushed a number of enemy patrols and columns, the enemy will resort to sending a number of foot soldiers ahead of the column to "sweep the path" by searching buildings and likely ambush spots for waiting guerrillas. Through this method, the enemy hopes to contact and discover the guerrillas before the main column has entered the ambush zone.

Since this method is time-consuming and slows the speed of the column tremendously, it will be resorted to only in areas where guerrilla activity has been frequent, or in the case of special convoys or to protect important potential targets. Nevertheless, it remains possible for the guerrillas to evade this "sweep" and successfully ambush the enemy.

This evasion depends on an increased number of lookouts or forward observers by the guerrillas. When the ambush is set up, the rear support teams assume their positions as for a normal ambush, and a number of forward lookouts take up positions from which they can watch the selected ambush site carefully. The guerrilla assault and close support teams, however, do not assume their ambush positions immediately. Instead, they remain hidden a short distance away.

As the enemy "sweep" moves through the ambush site, they are monitored by the forward observers. The enemy will thus enter the ambush zone, search it and, finding nothing suspicious, they will move on. Once they have passed the ambush site. the forward observers can report this fact to the waiting assault and close support teams, who can then carefully infiltrate into the site to occupy their ambush positions. The enemy force, thinking the ambush site to have been cleared by the sweep, thus falls into the guerrilla trap and is successfully ambushed.

Another tactic which is practiced by enemy troops who have

ENEMY "SWEEP" OPERATIONS

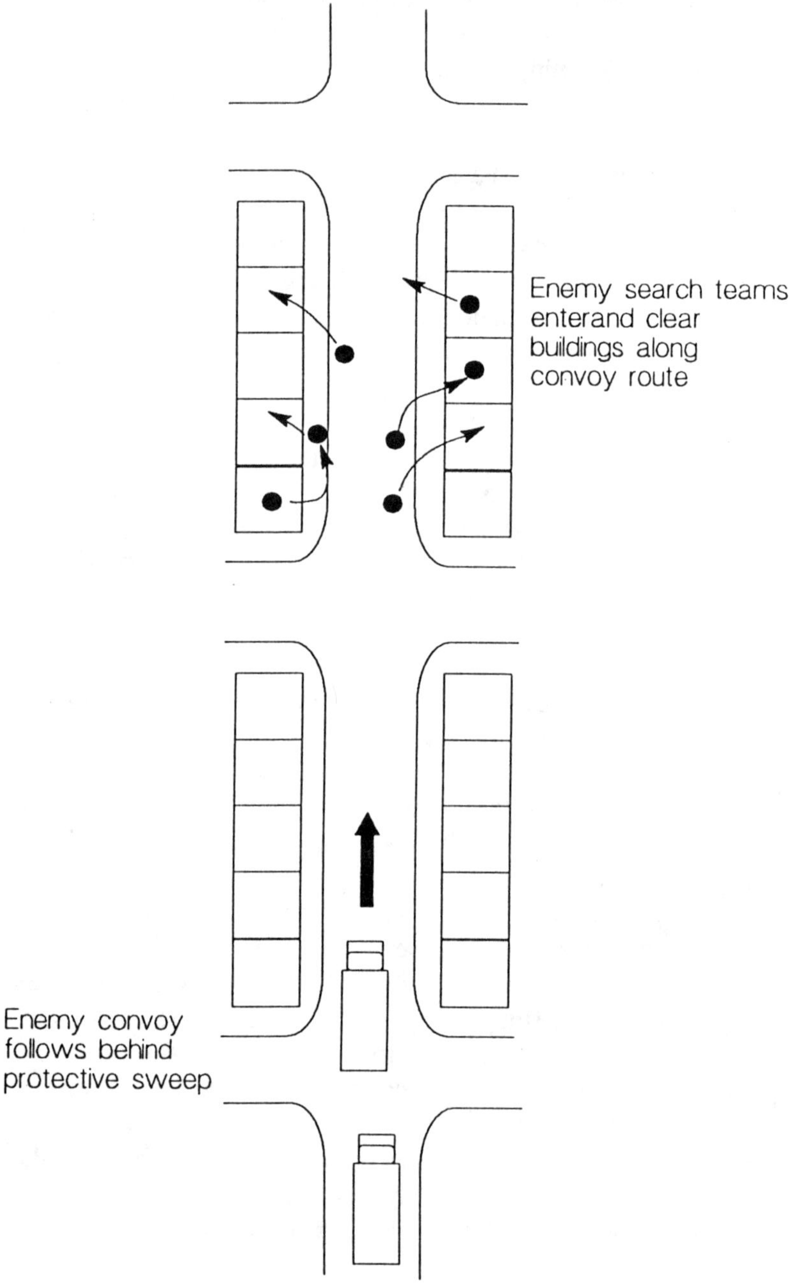

been ambushed successfully in the past is to accompany each column with a tank, APC, IFV or some other armored escort. These armored vehicles are capable of focusing an enormous amount of firepower on the ambushing guerrillas and must be eliminated as one of the actions of the ambush.

Tanks can be immobilized in urban areas and thus serve to "box in" the rest of the enemy force. Infantry anti-tank weapons such as the American LAW and the Soviet RPG-7 are incapable of penetrating the frontal armor of a Main Battle Tank or an Infantry Fighting Vehicle, but they can penetrate the rear or side armor of such a vehicle. In the urban ambush, guerrillas in high buildings armed with captured enemy weapons can destroy the armored vehicle by putting an RPG or rocket through the top, sides or rear armor.

If captured anti-tank weapons are not available to the guerrillas, they must use cruder methods of combatting the AFV. Home-made or improvised rocket launchers will probably not be able to penetrate the vehicle's armor, but they will be able to blow off a track or damage the engine, thus immobilizing the vehicle.

In the urban situation, the most lethal weapons of the armored vehicle is its machine gun.

To use the machine gun, however, the gunner must expose his head and shoulders, making him vulnerable to guerrilla snipers. The tank commander also usually rides with the hatch of the tank open for maximum visibility.

If the guerrilla lookouts report that the approaching enemy is escorted by tanks or IFVs, guerrilla sharpshooters must be nominated to specifically pick off the vehicles' commanders and machine gunners in the first burst of fire when the ambush is sprung.

Once the vehicle has been immobilized and disarmed, it can be destroyed by Molotov cocktails tossed into the hatch or onto the engine. Once on fire, its internal fuel and ammunition supply will detonate and destroy it.

Antitank mines are also capable of disabling or destroying armored vehicles, but are difficult to utilize in urban areas with paved roads.

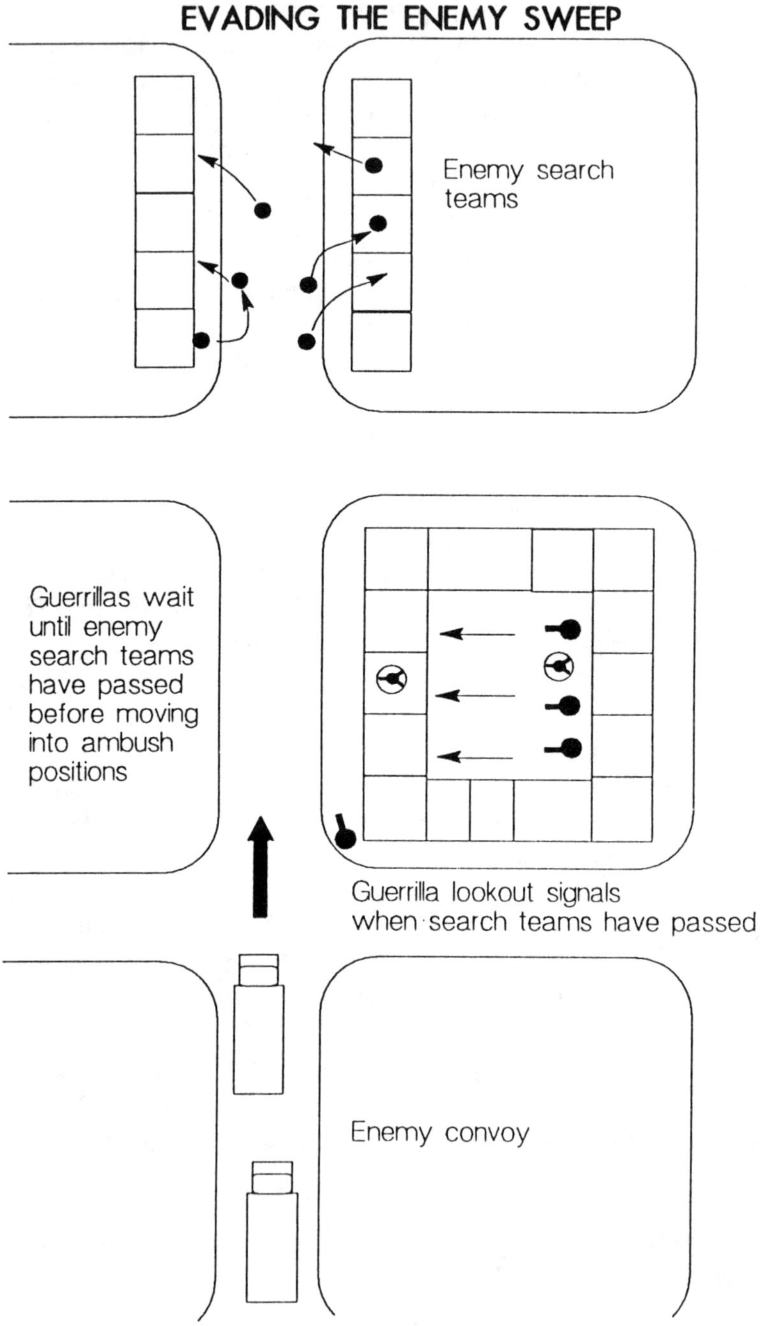
EVADING THE ENEMY SWEEP
Enemy search teams
Guerrillas wait until enemy search teams have passed before moving into ambush positions
Guerrilla lookout signals when search teams have passed
Enemy convoy

If the tank survives the initial burst of fire from the guerrillas by closing its hatch and employing its armored visors, it will remain impervious to small arms fire, but will suffer from severely restricted vision. The tank can then be destroyed by guerrilla assault teams who approach and attach explosive charges directly to it.

The only way to establish absolute security for the convoy is to station troops along both sides of the route for its entire length, a strategy known as "picketing." By thus occupying the entire convoy route, the enemy prevents guerrilla ambush teams from moving into position.

This strategy, however, causes still other problems for the enemy. The convoy route may be mined or booby-trapped beforehand, although this is difficult in an urban setting (since it is difficult to conceal mines in paved areas). Guerrillas may also hit and run from a distance using stand-off weapons such as RPGs, rocket launchers or grenade launchers.

The picket, moreover, requires a large number of enemy troops who are thus drawn from other duties. Under these circumstances, the best strategy for the guerrilla to follow is that of avoiding the enemy where he is strong and attacking him where he is weak. The guerrillas will thus ignore or merely harass the enemy supply convoy, and will attack in force at some other target which has been drained of its manpower. Thus, by using his advantage of local superiority of force, the urban guerrilla is able to strike at the enemy where he is least expected. The object is to keep the enemy constantly off balance and unable to concentrate his forces anywhere without inviting a guerrilla attack somewhere else.

In addition to defensive counter-ambush operations, the enemy will be forced to resort to offensive counter-insurgency operations to locate and annihilate the guerrillas.

In the absence of information about the precise location of the guerrillas, the enemy will resort to flooding suspected areas with troops in an attempt to flush them out and force them into open combat. This attempt most often takes the form of large-scale sweeps of suspected guerrilla-infested areas.

The sweep will most often take place early in the day (three or four in the morning) when people will be asleep. A force of enemy troops or state security forces will cordon off a section of the city with roadblocks, checkpoints and observation points. The outer ring of this net will cover all the major approaches to the cordoned area—bridges, highways, etc. It has as its goal the diversion of traffic from the area to be searched. The targeted area is thus cut off from contact with other parts of the city.

The inner security net will be escape-proof, and will cover every possible exit from the searched area—alleys, roads, footpaths. Its purpose is to prevent the escape of guerrillas who are caught in the sweep.

Once the security net has been established and the area has been cordoned off, state security forces will begin a systematic house to house search of the area. Beginning at one end, they will move through the area searching houses, detaining suspicious persons and annihilating any guerrillas encountered. These search teams are followed by infantry reserves who move up to wipe out any spots of resistance which are encountered. Helicopters and air observers will watch for fleeing guerrillas and direct enemy forces to them.

This process continues until the entire area has been "swept."

The primary defense against this sort of urban sweep is good intelligence and early-warning contacts. Such large-scale operations are impossible for the enemy to carry out quickly and with the element of surprise. Guerrilla contacts within the state security forces may be able to warn the network of the enemy's plans, giving the guerrillas time to remove supplies and personnel from any safehouses which are in the targeted area before the sweep takes place. The safehouse can, if necessary, be re-established once the enemy forces have left the area.

The effects of such sweeps can be minimized by widely dispersing the locations of the guerrilla safehouses and hideouts. Even if the sweep is not successfully avoided, the guerrillas will lose only one base of operations and thus minimize their losses.

If no early warning can be received before the sweep operation starts, the guerrillas must depend on their security network for a limited warning. Each safehouse maintains a network of security

ENEMY URBAN SWEEPS

State security forces set up net to cordon off area to be searched

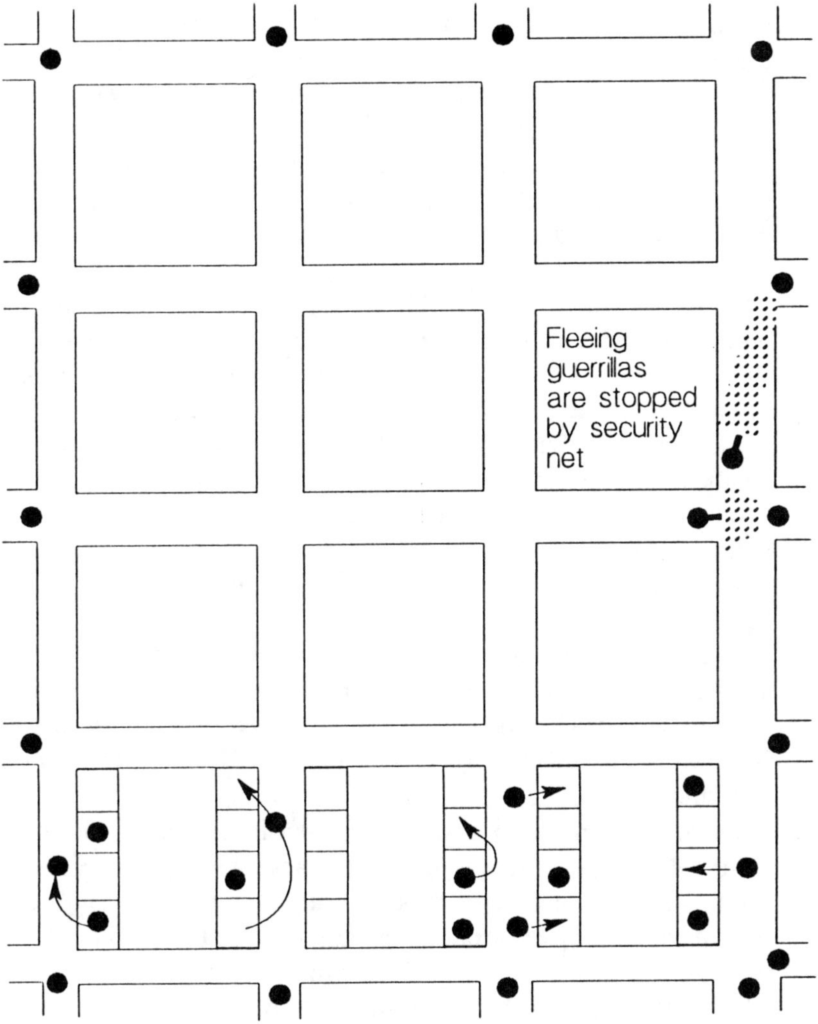

Search teams enter houses looking for guerrillas

and surveillance to watch for enemy attempts to surround the guerrillas. When the enemy cordon first begins, these security elements must inform the safehouses in the area, who thus have time to flee before the enemy security cordon is complete. Fighters and supplies may be exfiltrated from the area through the sewer or other subterranean systems, or they may be hidden from the enemy searchers in specially-constructed hiding places and caches.

Once the enemy cordon is complete, trapped guerrillas will have no choice but to hide from the enemy search teams or, using forged or stolen identity papers, attempt to bluff their way out as "innocent civilians." In the latter case, all incriminating evidence such as weapons, ammunition, maps, or any other object must be abandoned.

Guerrillas who choose to hide from enemy searchers must remain armed, and should hide in places which are mutually supportable by fire if they are discovered. If one guerrilla is discovered by the enemy, the rest should open fire to annihilate the search team, giving the guerrillas the opportunity to move into another hiding place.

Supplies that must be left behind in safehouses to facilitate escape should be booby trapped to slow enemy searchers and to destroy the safehouse, denying the enemy any potential intelligence information.

In practice, such large-scale sweeps never work as well as they are planned. Gaps will open between enemy search teams, and the cordon will never be absolutely leakproof. As a result, it is often possible for guerrillas to evade the search completely and emerge unscathed.

A much more dangerous form of counter insurgency is the use of so-called "hunter-killer" teams to track down guerillas and annihilate them. Each small, expertly-trained team is assigned the task of hunting down a particular guerrilla unit. In effect, the enemy attempts to use "counter-guerrillas" to combat the insurgents. Although this is largely a strategy to combat rural guerrilla bands, it can be applied to urban areas as well.

The hunter-killer teams must be combatted with strong security and intelligence measures. Guerrilla safehouses must be protected by a ring of early-warning and security operatives who keep a lookout for suspicious persons or suspected enemy surveillers. The local population must also serve as an early warning and intelligence network, since a great deal of the "hunter-killers'" tactic is to obtain information from the population concerning guerrilla operations. Security precautions such as compartmentalization and secure communications limits the amount of information which can be passed on to the enemy teams.

Once the location of the hunter-killer team has been established by the guerrillas, it is a simple matter to lure them into a counter-ambush.

Another tactic sometimes used by counter-insurgency forces is to form a Force X, a false guerrilla unit which roams the area, deceiving rebel sympathizers into giving them aid and information concerning the real guerrillas. This deception has as its goal the identification and elimination of the guerrilla infrastructure, cutting the guerrillas off from their civilian sources of supply, shelter and information.

Once again, the key to defeating this tactic is good security precautions. All efforts must be made to protect the guerrillas' contacts among the population, since they are every bit as important to the resistance as the armed fighters. Go-betweens and couriers must be used for secure contacts and exchanges. Passwords and ciphers must be jealously guarded and used to identify imposters. Security measures must be perfected and strictly observed.

The enemy may also resort to offering "rewards" for civilians who turn in guerrillas, or they may offer amnesty to guerrillas who surrender their weapons and turn themselves in. These deceptions have as their purpose that of gathering information and intelligence on the resistance network.

It is extremely difficult to defeat a true guerrilla force, one that has the support of the population and that successfully uses the principles of guerrilla warfare—mobility, surprise, local superiority of force, and knowledge of the local terrain. Guerrillas also possess a significant advantage over the state security forces—

while the security forces must win the conflict, the guerrillas need only avoid defeat.

7

SECURITY AND COMMUNICATIONS

The principles of clandestine security are fairly simple, but must be followed to the letter. The very existence of the guerrilla force depends on its ability to remain unseen and elusive, constantly hidden from the state security forces which are looking for it. Captured guerrillas can hope for little mercy; they will most likely be tortured for information and then publicly executed as a warning to others.

The organization of the guerrilla forces must be characterized by compartmentalization, with each section or division being isolated from the others. All information is exchanged on a strict "need to know" basis. No one should know anything more than that needed to complete his work.

The guerrilla forces should make no attempt to contact or work with prominent or outspoken critics of the regime. These persons will be under constant surveillance by the state security forces and will almost certainly be arrested for questioning. If they know nothing about the operations of the guerrillas, they can betray nothing.

The guerrillas are very dependent on secure communications to provide a flow of information and intelligence. Ambushes cannot be planned without carefully gathered information. Operations in an area cannot be coordinated unless different guerrilla units can communicate quickly and clandestinely. For these reasons, the resistance network must develop a secure method of passing information and intelligence.

Sensitive matters should never be discussed over the phone, as the state security forces will almost certainly have the telephone system tapped. No sensitive matters should be discussed in the safehouse, either, since, once they have discovered it, the security services will infiltrate it with listening devices and cameras.

Some state security forces have the use of laser devices which can pick up the vibrations of a closed window from considerable distances and decipher what is being said on the other side.

The best method of communication for the guerrillas are couriers and runners. These should be unsuspected persons such as children, women and senior citizens, who provoke little suspicion among the security forces. Sympathetic mail-carriers, priests or repair-persons can also serve as couriers, since they can move freely and visit many people without raising any suspicion. Guerrilla couriers can also use captured military or police uniforms. Long-distance couriers can pose as truckers, airline employees or train employees.

The courier system must be heavily compartmentalized to provide security. Messages to be communicated are given to a trusted intermediary, who then passes the message to a courier. The courier does not know what the message means, who it is from or who it is going to. His job is simply to transport it to another intermediary, who passes it on to the recipient. Thus, no single element in the chain, if captured by security forces, is able to compromise the entire network.

The system can be made even more secure by the use of "dead drops." Dead drops are nothing more than out-of-the-way spots where packages or envelopes can be safely stashed until they are picked up by an intermediary or courier. Suitable dead drops might include a hollow tree, a crack in a particular wall, under a particular rock or board, or any other location in which a package can be hidden safely. Some signal must be pre-arranged to let the courier know that there is a message in the dead drop. Suitable signals include a mark on a particular lamp post, a certain folded newspaper left on a park bench, a colored piece of paper left in a particular spot, or any other method which is noticeable but does not arouse suspicion.

The following example illustrates how this guerrilla communication network works: A resistance fighter prepares a message for transmittal and leaves it in a secure dead drop. The intermediary is informed by pre-arranged signal that a message awaits him, and he retrieves it from the dead drop. He leaves it in another dead drop, from which it is retrieved by another courier. This courier transports the message to the desired destination, where he leaves it in another dead drop. It is retrieved by another intermediary who leaves it in a final dead drop for delivery. The recipient receives it there.

Obviously, this method will take a great deal longer than a simpler method of communication, but its foolproof security makes it useful for high-level orders and sensitive intelligence data. Less secretive messages can be delivered simply by using a courier to transfer the message.

In any case, the message must be in a code or cipher so that the courier is unable to read it and so that it is useless to the state security forces if they intercept the message. Codes and ciphers used by guerrillas must fit a number of criteria. They must be simple to use and easy to decipher. Each ciphered message must have one, and only one, possible deciphered message. It must be secure enough to serve its purpose before it can be broken.

The simplest form of code or cipher is the single substitution cipher. In this method, the code letter is formed by a pre-arranged pattern of deviation from the message letter. For example, if the agreed-upon code is to use the second following letter of the alphabet, the code letters for SEND WEAPONS becomes UGPF YGCRQPU. For greater security, the cipher letters should not be divided according to the words formed, but should be grouped into uniform blocks of four or five letters. Thus, the message SEND WEAPONS would be ciphered as UGPFY GCRQP USJEX. The final letters in this message SJEX are fillers which have no meaning. They are only there to complete the block of five letters.

Single substitution ciphers are easy to break, however. They can be cracked using the knowledge that certain letters in the English language appear more frequently than others. By counting the number of times the code letter appears in the message, it is possible to assign frequencies to them which permit educated guesses as to which letter each code represents.

The best form of cipher is the so-called double substitution cipher, known to cryptologists as the Vigenere cipher. It makes use of an alphabetical table and a key word. The Vigenere code works like this:

Suppose the message to be sent is : NEED MONEY FOR RAID.

The agreed-upon key word is VICTORY. Begin the ciphering by repeating the key word under the message to be sent:

```
NEED MONEY FOR RAID
VICT ORYVI CTO RYVI
```

The simple Vigenere alphabetical table is shown on the next page. To find the proper code letter, read the message letter across the top of the Vigenere table and the key word letter from the side, and choose the cipher letter which appears at the intersection of these columns. In this example, the first message letter is N and the first key word letter is V. Looking at the table, these columns intersect at the cipher letter I. Continuing this process gives the cipher letters:

```
NEED MONEY FOR RAID
VICT ORYVI CTO RYVI
IMGW AFLZG HHF IYDL
```

Thus, the coded message to be sent is: IMGWA FLZGH HFIYD LAMRZ.

Once again, the final letters AMRZ are meaningless: they are tacked on to complete the final group of five letters. Notice that each of the three E letters in the message is now ciphered by a different code letter. This defeats the "letter frequency" method of decoding.

To decode the message, write the key word under the code letters:

```
IMGWA FLZGH HFIYD LAMRZ
VICTO RYVIC TORYV ICTOR
```

VIGENERE ALPHABET TABLE

	A	B	C	D	E	F	G	H	I	J	K	L	M	N	O	P	Q	R	S	T	U	V	W	X	Y	Z	
A	a	b	c	d	e	f	g	h	i	j	k	l	m	n	o	p	q	r	s	t	u	v	w	x	y	z	A
B	b	c	d	e	f	g	h	i	j	k	l	m	n	o	p	q	r	s	t	u	v	w	x	y	z	a	B
C	c	d	e	f	g	h	i	j	k	l	m	n	o	p	q	r	s	t	u	v	w	x	y	z	a	b	C
D	d	e	f	g	h	i	j	k	l	m	n	o	p	q	r	s	t	u	v	w	x	y	z	a	b	c	D
E	e	f	g	h	i	j	k	l	m	n	o	p	q	r	s	t	u	v	w	x	y	z	a	b	c	d	E
F	f	g	h	i	j	k	l	m	n	o	p	q	r	s	t	u	v	w	x	y	z	a	b	c	d	e	F
G	g	h	i	j	k	l	m	n	o	p	q	r	s	t	u	v	w	x	y	z	a	b	c	d	e	f	G
H	h	i	j	k	l	m	n	o	p	q	r	s	t	u	v	w	x	y	z	a	b	c	d	e	f	g	H
I	i	j	k	l	m	n	o	p	q	r	s	t	u	v	w	x	y	z	a	b	c	d	e	f	g	h	I
J	j	k	l	m	n	o	p	q	r	s	t	u	v	w	x	y	z	a	b	c	d	e	f	g	h	i	J
K	k	l	m	n	o	p	q	r	s	t	u	v	w	x	y	z	a	b	c	d	e	f	g	h	i	j	K
L	l	m	n	o	p	q	r	s	t	u	v	w	x	y	z	a	b	c	d	e	f	g	h	i	j	k	L
M	m	n	o	p	q	r	s	t	u	v	w	x	y	z	a	b	c	d	e	f	g	h	i	j	k	l	M
N	n	o	p	q	r	s	t	u	v	w	x	y	z	a	b	c	d	e	f	g	h	i	j	k	l	m	N
O	o	p	q	r	s	t	u	v	w	x	y	z	a	b	c	d	e	f	g	h	i	j	k	l	m	n	O
P	p	q	r	s	t	u	v	w	x	y	z	a	b	c	d	e	f	g	h	i	j	k	l	m	n	o	P
Q	q	r	s	t	u	v	w	x	y	z	a	b	c	d	e	f	g	h	i	j	k	l	m	n	o	p	Q
R	r	s	t	u	v	w	x	y	z	a	b	c	d	e	f	g	h	i	j	k	l	m	n	o	p	q	R
S	s	t	u	v	w	x	y	z	a	b	c	d	e	f	g	h	i	j	k	l	m	n	o	p	q	r	S
T	t	u	v	w	x	y	z	a	b	c	d	e	f	g	h	i	j	k	l	m	n	o	p	q	r	s	T
U	u	v	w	x	y	z	a	b	c	d	e	f	g	h	i	j	k	l	m	n	o	p	q	r	s	t	U
V	v	w	x	y	z	a	b	c	d	e	f	g	h	i	j	k	l	m	n	o	p	q	r	s	t	u	V
W	w	x	y	z	a	b	c	d	e	f	g	h	i	j	k	l	m	n	o	p	q	r	s	t	u	v	W
X	x	y	z	a	b	c	d	e	f	g	h	i	j	k	l	m	n	o	p	q	r	s	t	u	v	w	X
Y	y	z	a	b	c	d	e	f	g	h	i	j	k	l	m	n	o	p	q	r	s	t	u	v	w	x	Y
Z	z	a	b	c	d	e	f	g	h	i	j	k	l	m	n	o	p	q	r	s	t	u	v	w	x	y	Z
	A	B	C	D	E	F	G	H	I	J	K	L	M	N	O	P	Q	R	S	T	U	V	W	X	Y	Z	

Now, follow the key word letter across until you find the indicated cipher letter, then read the message letter from the top of the column. The first letter of the key word here is V. Reading across the V row, the cipher letter I is found in the N column, meaning the first letter of the message is N. Completing the process gives:

```
IMGWA  FLZGH  HFIYD  LAMRZ
VICTO  RYVIC  TORYV  ICTOR
NEEDM  ONEYF  ORRAI  DYTDI
```

The decoded message is NEED MONEY FOR RAID. The extra letters at the end are ignored.

This simple Vigenere system is not unbreakable: mathematical methods exist which allow security forces to determine the key word which is used to encipher the message and thus to break the code. These methods can be defeated, however, by making slight alterations in the system.

The table of the cipher, which is shown here with a standard alphabetical order in the rows, should be disjointed in an agreed-upon manner. This can be done by using a different order of letters; for instance, by using letters in the order in which they appear in a typewriter board:

```
q w e r t y u i o p a s d f g h j k l z x c v b n m
w e r t y u i o p a s d f g h j k l z x c v b n m q
e r t y u i o p a s d f g h j k l z x c v b n m q w
```

and so on to complete the 26 rows of the table. An even more secure method is to use a second code word to produce a disjointed alphabet in the body of the table. This code word is written at the beginning of the row, with repeated letters omitted, and the remaining letters in order after it. Thus, using the code word GREASE, the alphabetical rows would appear as:

```
g r e a s b c d f h i j k l m n o p q t u v w x y z
r e a s b c d f h i j k l m n o p q t u v w x y z g
e a s b c d f h i j k l m n o p q t u v w x y z g r
```

and so on.

To defeat the enemy's mathematical extraction of the initial key word used in ciphering, the guerrilla can use instead of a single word a long series of words, such as the text of a novel, a poem or story. It can be further agreed that the code word for disjointing the alphabetical table is a certain indicated word in the text.

For example, if we agree that the key text is that of the Declaration of Independence and that the code word for the body of the cipher is the fourth word in this text, or "course." The body of the table would look like this:

```
  A B C D E F G H I J K L M N O P Q R S T U V W X Y Z
A c o u r s e a b d f g h i j k l m n p q t v w x y z  A
B o u r s e a b d f g h i j k l m n p q t v w x y z c  B
```

and so on.

The message to be sent is printed and the text of the key text is printed under it:

```
SEND MONEY FOR WEAPONS
WHEN INTHE COU RSEOFHU
```

The Vigenere table is then consulted to produce the corresponding code letters.

This cipher system is absolutely secure unless it is betrayed by a defector or informer. It can serve for long messages which require utmost security.

For shorter messages with a lower degree of security, the guerrillas can use the number substitution cipher. This cipher uses the following alphabetical table:

	5 0	6 1	7 2	8 3	9 4
5 0	A	B	C	D	E
6 1	F	G	H	I	K
7 2	L	M	N	O	P
8 3	Q	R	S	T	U
9 4	V	W	X	Y	Z

Each letter of the message is, using this table, ciphered in a series of numbers. For instance, the message SEND WEAPONS can be ciphered into a series of numbers. The letter S, the first letter of the message, is found in the row numbered 8 and 3 and the column numbered 7 and 2. By placing the number of the row first and the column second, we can depict the position of S by the numbers 87, 82, 37 or 32. All of these numbers refer to the letter S, since it is found in row 8 column 7, row 8 column 2, row 3 column 7 or row 3 column 2. Thus, the same letter may be enciphered by up to four different numbers. The message SEND WEAPONS can be enciphered as 87-59-22-03 96-04-00-74-23-27-32 (breaking this into standard groups of five we get 87592 20396 04007 42327 32018) or as 32-04-77-08 41-59-50-29-78-72-87 (broken down as 32047 70841 59502 97872 87619).

To decode, simply break the number code into its digit pairs and find the letter corresponding to this row and column number.

Thus, the numbers 87 and 59 refer to row 8 column 7 and row 5 column 9. Using the table, we find these to be the letters S and E.

Notice that the letter J is missing from the table. This is to produce an even block of 25 letters in five rows of five. In the message to be ciphered, the letter I must be substituted for J.

To increase the security of this simple code, use an agreed-upon key word to produce a disjointed alphabet in the table of the number substitution cipher.

The cipher systems demonstrated here are capable of endless variations. They are not unbreakable, but they are not at all easy to break. For maximum security, the guerrilla force must periodically change its code words and key to prevent the enemy from eventually deciphering them. The greatest danger for any code is betrayal. If any guerrilla with access to the group's code is captured, deserts or turns informer, the code sequence must be immediately altered to insure continued security.

8

CONCLUSION

In order for the resistance network to be effective, it must begin its operations with a clear strategy in mind and a well thought-out method of implementing that strategy. This strategic focus is vital in urban guerrilla warfare. Urban partisan fighting is very different from urban warfare fought between regular armed forces. It is also very different in significant ways from guerrilla warfare fought by roving bands of rural fighters. Urban guerrilla warfare has a dynamic and intensity all its own.

It must be realized from the outset that a guerrilla force will not be able to topple a dictator or expel a foreign invader by itself. No matter how well-organized or well-trained, small bands of urban fighters will not be able to stand head-to-head with the trained professional armed forces of the dictator or the occupying power. Because of this, the guerrilla forces serve as a means of weakening and harassing the enemy, at the same time steadily building up its own organization, military strength and firepower. The goal of the guerrilla is to prove to the population that the regime is not invulnerable, and that by a concerted effort the people can topple the regime and expel the invader.

Thus, all guerrilla actions must have as their strategic goal that of building, organizing and arming a mass movement of resistance, an armed and organized population which is capable of fighting for its freedom.

The guerrilla's primary and initial aim is to weaken and demoralize the enemy and to seize from him the weapons, ammunition and supplies needed for the population to fight back. As more weapons are seized, through raids and ambushes, the guerrilla forces grow stronger, more numerous and better armed. As their strength grows, they are able to create zones or areas where the authority of the enemy government barely reaches, where enemy troops can move only during daylight and only in large, well-armed units. In these tiny "liberated zones," the resistance network itself becomes the de facto government.

In the early stages of the war, a liberated area may be held for only days or weeks at a time, until the enemy is able to concentrate enough force to overrun it. At this point, the guerrillas should not waste resources in a pitched battle to defend these territories. Rather, they should use their superior mobility to evade these attacks and return when the government troops have left.

As these liberated zones grow in space and endure in time, the armed forces of the resistance network will begin to approach the size and organization of a regular army which can stand in full battle against the enemy's forces. As the guerrilla forces in each city in the nation reach this point, planning can begin for the final blow, the armed mass uprising which will sweep the ruling regime from power. In one swift move, the armed population rises up across the nation to crush the enemy's military forces and sweep the oppressors from the country.

Such an armed uprising is only possible if the population is willing to play its part, to do what it can to defend the country. The population must embrace the motto adopted by fighters during the Spanish Civil War, "Better to die on your feet than to live on your knees."

How to Thrive on Rejection

A Manual for Survival

by Alan Abel

Illustrated by Simon Bond

AUTHOR OF 101 USES FOR A DEAD CAT

All of us have felt the painful pangs of rejection. After all, even presidents, millionaires, and beauty queens are sometimes rebuffed. Have you ever been denied a bank loan? A date? A job? A promotion? Alan Abel has, and over the past twenty-five years he has tested and perfected techniques for surviving rejection—indeed, for thriving on it.

Abel's solutions are inspired, creative, and zany, and they worked, at least for him. Here is his strategy for dealing with the Playboy organization, and with Simon and Schuster, and for persuading the Helmsley hotels to allow public access to their toilets.

Sometimes the best answer to rejection is a successful hoax, and Alan Abel has fooled the best of them, including Allen Funt (Abel fooled "Candid Camera"!) and *The New York Times*, which several years ago was tricked into running Abel's obituary.

So, here is Abel's manual for you to read and enjoy—and perhaps to apply, in your own battle for survival.

Only $5.95!